Fundamentals of Biophysics

Scrivener Publishing
100 Cummings Center, Suite 541J
Beverly, MA 01915-6106

Publishers at Scrivener
Martin Scrivener (martin@scrivenerpublishing.com)
Phillip Carmical (pcarmical@scrivenerpublishing.com)

Fundamentals of Biophysics

Andrey B. Rubin

Scrivener Publishing
WILEY

Co-published by John Wiley & Sons, Inc. Hoboken, New Jersey, and Scrivener Publishing LLC, Salem, Massachusetts.
Published simultaneously in Canada.

Cover design by Kris Hackerott

Library of Congress Cataloging-in-Publication Data:

ISBN 978-1-118-84245-4

Printed in the United States of America

10 9 8 7 6 5 4 3 2 1

Contents

Preface

It is generally accepted that in the twenty-first century biology will occupy one of the leading places in science. Biology will make an important contribution not only to our fundamental knowledge but also to the development of medicine, ecology, generation of new living species. Many problems in biology which are of interdisciplinary complex character lie at the boundaries between different sciences. To solve such problems it is necessary to apply ideas and methods of other exact sciences and first of all physics, mathematics, chemistry, and physical chemistry. It is by close cooperation between biology and other exact sciences that the development of biophysics as an independent boundary science has become possible.

In this book the main ideas of modern biophysics are presented in the form accessible to wide circles of readers. Biophysics (biological physics) is a science about physical and physico-chemical mechanisms of interactions which lie in the basis of biological processes. Physical properties of biopolymers and kinetics of cell metabolic reactions are responsible for molecular characteristics of biological processes. A biomacromolecule as the main structure element in a cell is considered in biophysics as a peculiar molecular machine where energy is transformed and conversed from one type of energy into another. It is pertinent to recall what Bruce Alberts, a well known American biologist, said about a cell. He wrote that "the entire cell can be viewed as a factory that contains an elaborate network of interlocking assembly lines, each of which is composed of a set of large protein machines" (Cell, 1998, vol. 92, pp. 291-294).

The real understanding of how these protein machines operate demands the knowledge of not only atomic equilibrium structure but also our understanding of kinetic and energy characteristics of intermediate transformations. In the post-genome sequencing era the first priority is given to the mechanisms of intramolecular mobility of macromolecular complexes as the base of their activity. Such an approach corresponds to the biophysical concept of directed electron-conformational interactions when energy transformation and reaction product generation become a result of internal interaction between separate parts within the whole macromolecular complex. In other words, this is the concept of a "physical machine" put forward in 1970-1980s by D.S.Chernavsky, L.A.Blumenfeld, and M.V.Volkelshtein.

In theoretical biophysics, generalized kinetic and physical models of interactions allow to describe different biological phenomena. However the analysis of such models clearly demonstrates that different biological processes can very often be similar with respect to their molecular mechanisms. For example, mechanisms of primary photobiological processes (photosynthesis, visual reception), enzyme catalysis in the enzyme active center, and ion transfer through membrane channels are governed by similar physical principles. It follows that educational programs on biology at universities should necessarily include ideas of physics, mathematics and physical chemistry, thus illustrating their efficiency in solving biological problems. Biophysics bears the main responsibility to show an important role of regular application of ideas from exact sciences in studying biological processes.

The great experience gained at Moscow State University has shown that this requirement is satisfied when teaching of biophysics is carried on within a logically combined independent general course on biophysics. Such a lecture course on biophysics is given both to biology and physics undergraduates as well as to bachelors of medicine and agriculture. It is useful for scientists working in other fields who want to study biological problems using methods of exact sciences.

In this book the main ideas and concepts of theoretical biophysics (kinetics and thermodynamics of biological processes, molecular biophysics) are presented in a concise and simple form. It makes the base for the biophysical analysis of some important processes (membrane processes, primary photobiological processes and enzyme reactions) analyzed in the book. The material is quite understandable and accessible to specialists of different profiles and background. It is useful for the subsequent profound studying of biophysics using special literature sources for the application of the biophysical approach in different fields of biology.

1

Dynamic Properties of Biological Processes

Biological Kinetics. Intricate network of various reactions, specifically organized in time and space, underlie both cell exchange processes with the environment and internal metabolism. In biological systems, components interact continuously with each other, which for the most part specifies the nature of dynamic behavior of intact biological systems, mechanisms of their self-control and governing named kinetics of biological processes. As a result of such processes, concentrations of different substances, the number of individual cells and the biomass of organisms change; the other values may also vary, for instance the transmembrane potential in the cell. Upon description of the kinetics in biological systems, the basic initial prerequisites are generally the same as in chemical kinetics.

It is believed that changes of variables at every time moment can be described using corresponding differential equations. In addition to variable values, a kinetic system has a set of specific parameters that remain unchanged during its examination and characterize the conditions of reactions (temperature, humidity, pH, and electrostatic conductivity). As a rule, the constant values of the reaction rates are determined by such parameters.

Let us analyze an elementary example of closed cell population in which multiplication and death occur concurrently and which are abundant in nutrients. The questions arise: how is the number of cells changed in such a system with time, and can a stationary state eventually form in it when the number of cells remains the same? This kinetic problem may be solved with the use of differential equations. Let at moment t the concentration of cells in the environment be N. The rate of the cell concentration changes in the environment $\frac{dN}{dt}$ is the net sum of their multiplication rate (v_{mitipl}) and death rate (v_{death}).

$$\frac{dN}{dt} = v_{multipl} - v_{death}.$$

In an ordinary case, the multiplication rate, which is the increase in the cell concentration per time unit, is proportional to their number at every moment, i.e.

$$V_{multipl} = k_1 N,$$

where k_1 is the proportionality constant dependent on the environmental conditions (temperature, the presence of nutrients, etc.)

Correspondingly

$$V_{death} = k_2 N,$$

where k_2 is the constant determining the intensity of processes of cell death. Hence it follows that

$$\frac{dN}{dt} = k_1 N - k_2 N = kN, \tag{1.1}$$

where $k = k_1 - k_2$.

By solving the above equation we will see how the cell concentration is changed with time in the environment $N = N(t)$. By integrating Eq. (1.1) we get

$$\int \frac{dN}{N} = \int k dt;$$

$$\ln N = kt + c;$$

$$N = N_0 e^{kt}, \tag{1.2}$$

where N_0 is the cell concentration at zero time $t = 0$ of the examining the system.

It can be seen that depending on the ratio of the death rate constant (k_2) and multiplication rate constant (k_1) the destiny of this closed population will be different. If $k_1 > k_2$, $k > 0$, the system will give rise to the unlimited growth of the cell number.

$N(t) \to \infty$ at $t \to \infty$.

If $k_1 < k_2$, the population will dye out with time

$N(t) \to \infty$ at $t \to \infty$.

And only in a particular case when $k_1 = k_2$ the number of cells will remain constant

$N = N_0$.

Another example of the model of the population growth in the environment with a limited amount of nutrients is the known equation of a logistic curve. The Verhulst equation is as follows

$$\frac{dN}{dt} = k_1 N \left(\frac{N_{\max} - N}{N_{\max}} \right). \tag{1.3}$$

Here $N_{\max}$ is the maximal population number possible under such conditions. Curve $N = N(t)$ described by the above equation is shown in Fig. 1.1. At the initial period of growth, when $N \ll N_{\max}$ the curve is exponential. Then, after the inflection, the slope gradually decreases and the curve approaches the upper asymptote $N = N_{\max}$, i.e. the maximal attainable level under such conditions.

But as compared to the typical chemical kinetics, the biological kinetics has the following specificity.

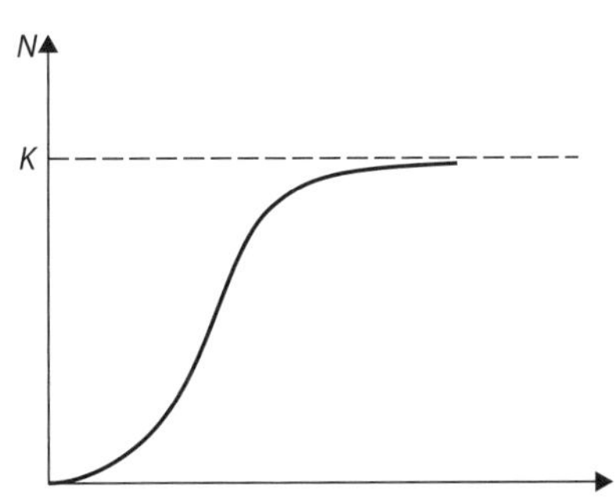

Figure 1.1 Logistic curve.

1. The variables can be not only substance concentrations but also other values.
2. The variables change not only in time but also in space (diffusion of reagents across biomembranes).
3. A biological system is heterogeneous in space and the conditions of the reagent interaction can vary in different sites of the system).
4. There exist special mechanisms of self-regulation functioning by the feedback principle.
5. The power of the polynomial in the right-hand side of the differential equation is not always dependent on the reaction order.

As a rule, a hydrodynamic model of a vessel connected with liquid inflow and outflow fluxes passing simultaneously, is taken as a simple model of an open system. The liquid levels in the vessel are dependent on the rates of liquid inflow and outflow fluxes. When the rates are the same, the liquid level would remain constant and a stationary state forms in the system. A change in the rate of at least one flow would cause a corresponding shift in the stationary level of the liquid in the vessel.

The Feedback Principle. Let us provide our hydrodynamic model with a special device that can increase or decrease the rate of liquid outflow upon rotation of the faucet at the outlet of the vessel depending on the liquid level changes. Such a system is shown in Figure 1.2. An electromotor rotates the stopcock according to the signal received from the photocell. The electric current generated in the photocell is governed by the level

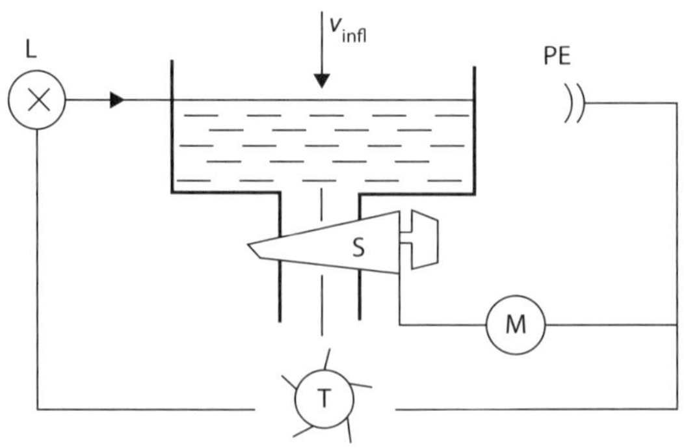

Figure 1.2 Hydrodynamic model of a feedback system. L, lamp; PE, photocell; S, stopcock; M, electromotor; T, turbine.

of light absorption that changes parallel to the liquid level in the vessel. The photocell lamp and the electromotor are power supplied from a small turbine whose blades are rotated by the water outflow. In this model the feedback principle maintains to a certain extent the liquid level upon varying the water inflow as a result of self-regulation. In biological systems, the feedback principle is used to regulate many enzyme reactions where the activity of enzymes varies depending on the reagent concentration or external conditions. As a result the concentration of reaction products remains constant. Biological systems may have different stationary regimes depending on the values of controlling parameters. It is also possible that fluctuations at the stationary states appear when concentrations of intermediate substances change in time at regular intervals and fixed frequency. Finally, under certain combinations of chemical reactions and diffusion processes proceeding concurrently, a special type of three-dimensional structure may appear in the originally homogeneous system.

The cardinal problem for biophysicists is to obtain characteristics of various dynamic regimes in complicated systems and to study conditions and parameters when they are realized in living cells. This can be done by studying the properties of stationary regimes, their stability and transition to a stationary state.

Elementary Model of Open System. Let us analyze an elementary model of an open system which exchanges substances a and b with the environment, coupled to the reversible first order reaction of transformations $a \rightleftarrows b$. In Fig.1.3, a and b are concentration variables within the system; A and B are constant concentrations of the same substances in the external vessels; and k_1, k_{+2}, k_{-2}, and k_3 are rate constants of the processes.

Though being very simple, the model reflects the basic features of metabolic processes in a cell. The supply of the substrate and the release of metabolites into the environment are triggered by the reactions $A \rightarrow a$, $b \rightarrow B$, and the transformation $a \rightleftarrows b$ corresponds to the processes of cell

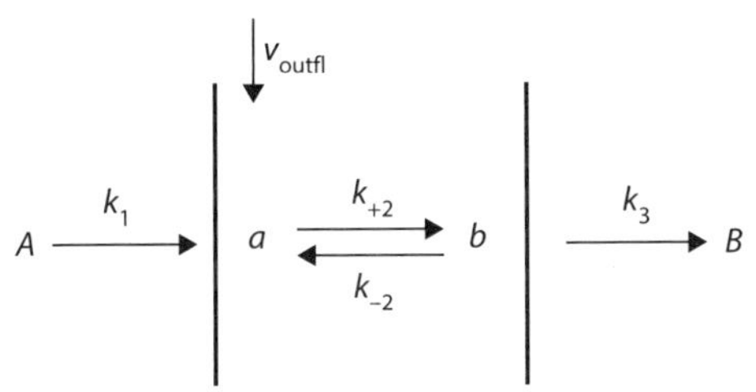

Figure 1.3 Open system: a model of metabolic processes in a cell.

metabolism. For example, glucose and oxygen as substrates for respiration are supplied to a cell at stage $A \rightarrow a$; stage $b \rightarrow B$ corresponds to the release of CO_2 and H_2O to the outside of the cell, and the entire metabolic respiratory cycle of the glucose molecule transformation can be described by reaction $a \rightleftarrows b$. The rate constant values are naturally phenomenological and generalized and therefore cannot be attributed to any particular biochemical stage. However, as we have seen, even such an utterly simplified model reveals basic features of a combination of metabolic reactions of a cell as an open system.

The kinetic equations for this system are as follows:

$$\frac{da}{dt} = k_1(A-a) + k_{-2}b - k_{+2}a = f_1(a,b),$$

$$\frac{db}{dt} = k_{+2}a - k_{-2}b - k_3(b-B) = f_2(a,b). \tag{1.4}$$

Inasmuch as at a stationary state, variables (a, b) are constant, then

$$\frac{da}{dt} = 0 \text{ and } \frac{db}{dt} = 0.$$

Let us equate the right-hand side of eq. (1.4) to zero:

$$\frac{da}{dt} = f_1(a,b) = 0; \ \frac{db}{dt} = f_2(a,b) = 0. \tag{1.5}$$

We get the system of algebraic equations:

$$\begin{cases} -(k_1 + k_{+2})a + k_{-2}b + k_1 A = 0, \\ k_{+2}a - (k_{-2} + k_3)b + k_3 B = 0, \end{cases} \tag{1.6}$$

from which stationary values $\bar{a}$ and $\bar{b}$ are known:

$$\bar{a} = \frac{Ak_1(k_{-2} + k_3) + k_3 k_{-2} B}{k_1 k_{-2} + k_1 k_3 + k_{+2} k_3}; \ \bar{b} = \frac{k_3(k_1 + k_{+2})B + k_1 k_{+2} A}{k_1 k_{-2} + k_1 k_3 + k_{+2} k_3}. \tag{1.7}$$

The above values do not depend on the initial conditions, i.e. on the initial values $a = a_o$ and $b = b_o$ at time $t = 0$ but depend only on the constant values and substance concentrations in the external vessels *A* and *B*. This means that whatever the initial state of the system is, the only stationary regime with $a = \bar{a}$, $b = \bar{b}$ will be finally formed. The system of differential equations (1.4) is solvable if dependences $a = a(t)$ and $b = b(t)$ are defined explicitly and it is ascertained how variable concentrations change with time. The solution will be

$$a(t) = c_1 e^{\lambda_1 t} + c_2 e^{\lambda_2 t},$$

$$a(t) = c_1 \chi_1 e^{\lambda_1 t} + c_2 \chi_2 e^{\lambda_2 t}, \qquad (1.8)$$

where c_1 and c_2 are constants dependent on the initial conditions.

At $t = 0$, $a_o = c_1 + c_2$, $b_o = c_1 \chi_2 + c_2 \chi_2$, i.e.

$$c_1 = \frac{a_o \chi_2 - b_o}{\chi_2 - \chi_1}, \qquad (1.9)$$

$$c_2 = \frac{b_o - a_o \chi_2}{\chi_2 - \chi_1}.$$

The values λ_1 and λ_2 are exponential indices determined from the characteristic equation

$$\begin{vmatrix} -(k_1 + k_{+2}) - \lambda & k_{-2} \\ k_{+2} & -(k_{-2} + k_3) - \lambda \end{vmatrix} = 0,$$

or $\lambda^2 + \lambda(k_1 + k_{+2} + k_{-2} + k_3) + (k_1 k_{-2} + k_1 k_3 + k_{+2} k_3) = 0$.

The values χ_1 and χ_2 are estimated from the equation for the distribution coefficients

$$k_{-2}\chi^2 + (-k_1 - k_{+2} + k_{-2} + k_3)\chi - k_{+2} = 0.$$

If $t \to \infty$, then $e^{-\lambda_1 t} \to 0, e^{-\lambda_2 t} \to 0$, i.e. $a \to \bar{a}$ and $b \to \bar{b}$. This shows that finally the system attains a stationary state independent of the initial conditions $a = a_o$ and $b = b_o$. This is the so-called property of equifinality of stationary states that is inherent to open systems and frequently observed in biological processes. Although the initial conditions do not affect the

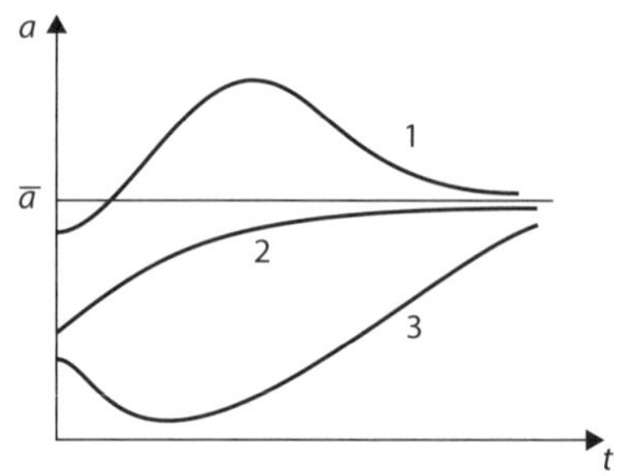

Figure 1.4 Transition curves $a(t)$ for scheme 2: 1, overshoot; 2, monotonous; 3, false start.

values of a and b, they determine a specific shape of the curves of changes in $a(t)$ and $b(t)$ and the kinetics of the transition of the system from the starting point $a = a_0$, $b = b_0$ at $t = 0$ to the stationary state $a = \bar{a}$, $b = \bar{b}$ at $t \to \infty$. Figure 1.4 demonstrates several types of such transition curves $a(t)$.

Similar shapes of the curves were traced, for example, in physiological studies of the respiration rate under different initial conditions. It should be understood that the shape of the curves depends on the initial conditions and constant k_1, k_{+2}, k_{-2}, k_3, A, and B values and can be changed due to their combinations. Examination of a simple system (1.4) shows that analytical solutions are rather bulky and are specified by a great number of parameters. It is clear that at a great number of variables, it is not only difficult to obtain such solutions but also to determine how the kinetic behavior of the system depends on its parameters. Let us pay attention to the fact that equations (1.4) have only linear members in their right-hand sides with unknown variables to the first power. However as a rule in biological systems, many processes are non-linear. Thus, the rate of the simplest second-order bimolecular reaction is described mathematically as the product of reagent concentrations, i.e. in a model of such reactions the right-hand sides of equations include non-linear members. In this case, the search for the accurate analytical solutions runs into serious mathematical difficulties and at times is not possible at all.

Qualitative Analysis of the Model. The basic approach in the contemporary kinetics and mathematical modeling of biological processes is to refuse the search for exact analytical solutions of differential equations. The idea is to get qualitative characteristics of the dynamic behavior of the system: its various stable and unstable stationary states, transitions between them, oscillatory regimes, qualitative dependence of the system behavior on critical values of the parameters. Many of these problems may be solved using methods of the qualitative theory of differential equations which allow disclosing essential general properties of the model without explicit calculation of the unknown functions. The most essential feature of a stationary state is its stability. The stability of the system depends on its ability to return spontaneously to the initial stationary state after the introduction of external disturbances that deviate the system from the initial stationary point.

Stationary Point Stability. Let us take the simplest open system $A \xrightarrow{v_{\text{infl}}} |a| \xrightarrow{v_{\text{outfl}}}$, to which substance a is supplied from an external source at a constant rate $v_o = v_{\text{infl}} = \text{const}$.

The kinetic equation is

$$\frac{da}{dt} = v_{\text{infl}} - v_{\text{outfl}} = v_0 - ka = f(a)\,, \tag{1.11}$$

where k is the rate constant v_{outfl}. The solution of this equation is very simple. But let us determine the value of stationary point $a = \bar{a}$ and estimate its stability graphically without considering the analytical solution of equation (1.11). It is evident that the stationary state $\frac{da}{dt} = 0$ in eq. (1.11) is established at such a value of $a = \bar{a}$ when the inflow and outflow rates become equal ($v_{\text{infl}} = v_{\text{outfl}}$).

Figure 1.5 shows the dependence of rates $v_{\text{infl}} = v_o$ and $v_{\text{outfl}} = ka$ on the a value. The plots of v_{infl} and v_{outfl} are straight lines intersecting at the point where $v_{\text{infl}} = v_{\text{outfl}}$, i.e. at the stationary point $a = \bar{a}$.

Let us see whether this point is a stationary one. Assume that incident disturbance occurred in the stationary-state system $a = \bar{a}$ that results in

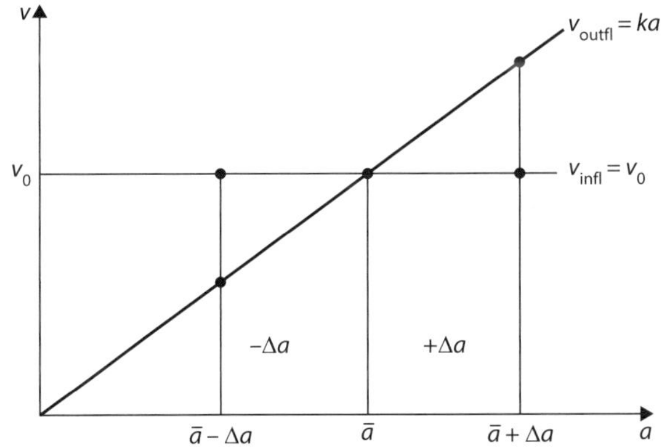

Figure 1.5 Dependence of inflow (v_{infl}) and outflow (v_{outfl}) rates on value a.

raising the stationary concentration of a by Δa and thus deviating the system from point $\bar{a}$ to the nearby point $\bar{a} + \Delta a$. In the new perturbed state where $a = \bar{a} + \Delta a$, the v_{infl} and v_{outfl} values are not equal any longer and consequently the concentration of $\bar{a} + \Delta a$ begins changing. As seen from the figure, at point $\bar{a} + \Delta a$ the ratio of v_{infl} and v_{outfl} is such that $v_{outfl} > v_o$. Thus the concentration of $a(\bar{a}+\Delta a)$ in the system should decrease spontaneously along with the inflow rate $v_{infl} = ka$ until the rates v_{infl} and v_{outfl} become equal again when $a = \bar{a}$.

It is obvious that a similar situation will be if the incidental shift from point $a = \bar{a}$ would result in a decrease in stationary concentration a that causes the transition of the system to point $\bar{a} - \Delta a$. As far as in this case $v_{infl} = v_o$, the $\bar{a} - \Delta a$ value and the outflow rate would grow until v_{outfl} becomes equal to v_o so that the system returns to the initial state $a = \bar{a}$. So, accidental deviations from the stationary point ($\pm\Delta a$) are compensated by the system itself which just means that the system is in a stable stationary state.

The Criterion of Stability. There is an easy analytical method for determining the stationary state stability, and we will use it to examine models of biological processes with two equations.

Let us take an elementary mathematical model with a simple first-order differential equation:

$$\frac{da}{dt} = f(a), \tag{1.12}$$

where the right-hand side $f(a) = v_{infl} - v_{outfl}$ may have different form. Let the system have a stationary point $a = \bar{a}$, where $\left.\frac{da}{dt}\right|_{a=\bar{a}} = 0$ and as a result $f(\bar{a}) = 0$. We will see how the stability of point $a = \bar{a}$ is associated with the properties of function $f(a)$. We set deflection $\xi = \Delta a$ close to the stationary point $a = \bar{a}$ so that

$$\xi = a - \bar{a} \tag{1.13}$$

is a low value

$$\left|\frac{\xi}{\bar{a}}\right| = \left|\frac{a - \bar{a}}{\bar{a}}\right| << 1.$$

Let us substitute $a = \bar{a} + \xi$ from eq. (1.13) into eq. (1.12). We obtain:

$$\frac{d(\bar{a}+\xi)}{dt} = \frac{d\xi}{dt} = f(\bar{a}+\xi). \tag{1.14}$$

Then we expand function $f(\bar{a}+\xi)$ in the Taylor series near point $\bar{a}$:

$$\frac{d\xi}{dt} = f(\bar{a}+\xi) = f(\bar{a}) + \left(\frac{df}{da}\right)_{a=\bar{a}} \xi + \frac{1}{2}\left(\frac{d^2 f}{da^2}\right)_{a=\bar{a}} \xi^2 + \dots \quad (1.15)$$

At the stationary point $f(\bar{a}) = 0$, so restricting ourselves to the first-order values, we obtain

$$\frac{d\xi}{dt} = f'(a)\big|_{\bar{a}}\, \xi . \quad (1.16)$$

This equation determines the time-dependent behavior of disturbance ξ introduced near the stationary point. From eq. (1.16) it follows that the change of $\xi = \xi(t)$ can be written as

$$\xi = \xi_0 e^{f_a'(\bar{a})t}, \quad (1.17)$$

where ξ_0 is the magnitude of the initial deviation at $t = 0$. It is clear that if $f_a'(\bar{a}) < 0$, then $\xi \to \infty$, at $t \to \infty$, that is the deviation disappear with time so that the system returns to the stationary point $\bar{a}$ which accordingly is stable. On the contrary, at the unstable state $\bar{a}$ the magnitude of the initial deviation increases with time as determined by the inequality $f_a'(\bar{a}) > 0$. In this case, it follows from eq. (1.17) that $\xi \to \infty$ at $t \to \infty$. So, at the stationary point the sign of the derivative of the right-hand side of the differential equation shows the type of stability of the stationary state. It may be easily checked that for system (1.11) $f_a'(\bar{a}) = -k < 0$ at point $\bar{a}$, i.e. at any $a = \bar{a}$ the stationary state is stable.

Several Stationary Points. Reactions of the second and higher orders can occur in a complicated system. In this case, equation $f(\bar{a}) = 0$ for determining the stationary point coordinates will have several roots, which is consistent with the fact that our system has several stationary states.

Let us analyze the model of a flow-through cultivator where concentration c is changed due to bacterial cell reproduction (γc^2), their death ($-\beta c$) and the adjusted constant rate v_0 of inflow from the outside that can be altered when required. The model is as follows:

$$\frac{dc}{dt} = v_0 - \beta c + \gamma c^2 = f(c). \quad (1.18)$$

For simplicity, we take $\gamma = 1$. By equating $f(\overline{c})$ to zero, we find that there are two stationary points in the system:

$$\overline{c}_1 = \frac{\beta}{2} + \sqrt{\frac{\beta^2}{4} - v_0} \quad \text{and}$$

$$\overline{c}_2 = \frac{\beta}{2} - \sqrt{\frac{\beta^2}{4} - v_0} \qquad (1.19)$$

Point $\overline{c}_1$ is not stable since the following equation is valid:

$$f'_{\overline{c}_1}(c) = 2\overline{c}_1 - \beta = 2\sqrt{\frac{\beta^2}{4} - v_0} > 0,$$

while on the contrary, point $\overline{c}_2$ is stable:

$$f'_{\overline{c}_2}(c) = 2\overline{c}_2 - \beta = -2\sqrt{\frac{\beta^2}{4} - v_0} < 0.$$

It is remarkable that the number of stationary points depends on the magnitude of v_0. As seen from eq. (1.19) while at $v_0 < \frac{\beta^2}{4}$ two stationary states $\overline{c}_1$ and $\overline{c}_2$ exist, at $v_0 > \frac{\beta^2}{4}$ there can be only one stationary state $\overline{c}_1 = \overline{c}_2 = \frac{\beta}{2}$. At $v_0 > \frac{\beta^2}{4}$ there are no stationary states at all, because $\overline{c}_1$ and $\overline{c}_2$ turn to be imaginary numbers which is not possible. So, by varying v_0 we alter not only the coordinates of the stationary points, but can also change their number. The value of $v_0 = \frac{\beta^2}{4}$, such that the number of stationary states changes or, as shown below, the type of their stability changes, is called a bifurcational value. In this case the v_0 parameter is a controlling one.

Reduction of the Number of Equations. As seen, some essential properties of stationary states can be revealed by studying the properties of the right-hand sides of differential equations without obtaining their accurate analytical solution. However such an approach yields good results when the models examined consist of a small number of equations (mostly of two equations). Obviously if it is necessary to take into account all variable concentrations of the intermediates involved even in simple biochemical cycles, the number of equations in the model would grow quite enormously. That is why for a successful analysis it will be necessary to decrease the number of equations in the initial model and to reduce the latter to a model consisting of a small number of equations that would nonetheless reveal the most essential dynamic properties of the system. Such a reduction in the number of equations cannot be made arbitrarily and hence accomplishment should obey the objective laws and rules. Otherwise it is hazardous to lose some important properties that will make the model inadequate to the simulated biological system.

Fast and Slow Variables. Reduction of the number of equations is based on the bottleneck principle or the principle of separation of variables into fast and slow ones in complicated systems. Let us analyze the essence of this principle. The heterogeneous nature of the structure of biological systems is revealed both structurally and dynamically. Different functional processes, individual metabolic cycles differ from each other significantly in their characteristic times (τ) and rates. Integrated biological systems are characterized by fast processes of enzyme catalysis ($\tau \sim 10^{-1}$–10^{6}s), physiological adaptation (τ from s to min), and reproduction (τ from several minutes and higher). Even within one individual sequence of interrelated reactions there are always the slowest and fastest stages. To be exact, this is the base for the bottleneck principle according to which the total rate of the substance transformation via the sequence of reactions is determined by the slowest stage (the bottleneck). This slow stage has the highest characteristic time (the lowest rate) as compared to all characteristic times of other individual stages. The total time of the process is in effect compatible with the specific time of the bottleneck. It is the slowest link that is principal as a controlling one since the action caused on it rather than on faster stages influences the overall rate of the processes. Thus, though complicated biological processes do include a great number of intermediate states, their dynamic characteristics depend on a relatively small number of individual slowest links. This shows that studies may be performed on models containing a remarkably less number of equations. The slowest stages correspond to slowly changing variables and the fastest stages to fast changing ones. The deep meaning of this is that when such a system is affected in any way (some disturbance is induced), in response all variable concentrations of interacting substances would begin changing also. But for different substances their changes would proceed at essentially different rates. In stable systems, fast variables will deviate rapidly, and then will rapidly return to their initial values. To the contrary, slow variables change for a long time during transition, and thus it determines the overall dynamic rate of the system. In real conditions, the system experiences 'pushes' from the outside which lead to visible changes in slow variables, but fast variables will remain near their stationary values. Then for fast variables, algebraic equations that determine stationary values of fast variables may be introduced instead of differential equations. This is the way to reduce the number of differential equations in the total system which will now include only slow time-dependent variables.

Let us consider two differential equations for two variables x and y such that

$$\frac{dx}{dt} = AF(x, y),$$

$$\frac{dy}{dt} = Q(x, y), \tag{1.20}$$

where $A \gg 1$ is a large value. This means that the product $AF(x,y)$ is a high value, and so the rate of changing $\frac{dx}{dt}$ is also large. Consequently, x is a fast variable. Let us designate and divide the right-hand and left-hand sides of the first equation by A. We get

$$\varepsilon\frac{dx}{dt} = F(x, y),$$

$$\frac{dy}{dt} = Q(x, y), \tag{1.21}$$

where $\varepsilon \ll 1$ ($\varepsilon = 1/A$).

At $\varepsilon \to 0$

$$\varepsilon\frac{dx}{dt} = F(x, y) = 0\cdot$$

That is why the differential equation for variable x may be replaced by the algebraic equation

$$F(x,y) = 0,$$

in which x is a stationary value dependent on y as a parameter, i.e. $x = x(y)$. In this regard, slow variable y is a governing parameter, so that by changing

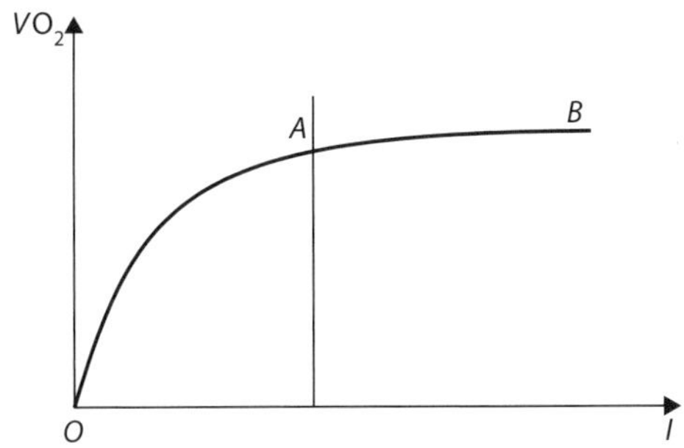

Figure 1.6 Dependence of oxygen release (v_{O_2}) versus the luminescence intensity (I) during photosynthesis.

the latter we can alter the coordinates of stationary point $x(y)$. In the above example (1.18) of a flow-through cultivator, value v_o (the rate of the cell inflow) played the role of such a governing parameter. By slowly changing this value each time we induced a relatively rapid adjustment of the stationary concentration of cells in the system (c is a fast variable). The addition of another equation describing this slower time-dependent change of v_o to cq. (1.18) could result in a complete description of the system with account of the fast (c) and slow (v_o) variables.

Within the same biological system, dependent on the external conditions, different chain links can play the bottleneck role. Let us examine, for example, the character of light curves of photosynthesis, i.e. the dependence of the oxygen evolution rate versus the illumination intensity (I) (Fig. 1.6). In region OA of this curve, at insufficient illumination it is the initial photochemical stages of light absorption and light-energy transformation in the pigment apparatus that become the bottleneck of the total photosynthetic O_2 release. It is worth mentioning that these processes by themselves are almost independent of temperature. This is the reason why at low illumination intensities the total rate of photosynthesis, or the rate of O_2 release, is known to vary very slightly with temperature within the physiological range (+5 to +30°). In this part of the light curve, the role of a fast variable is played by dark electron transfer processes that swiftly respond to any changes in light intensities and accordingly in the electron flux rate from reaction centers of the photosynthetic apparatus at low illumination.

But at higher light intensities, dark electron transfer and water splitting processes become the limiting stage on the AB part of the light curve. Under such conditions, dark processes become the bottleneck at large I values. They cannot consume the powerful electron flux from the pigment apparatus at high illumination which results in light saturation of photosynthesis. At this stage, because of the enzyme nature of dark processes, an increase in temperature evokes their acceleration and thus increases the total rate of photosynthesis (oxygen release) under conditions of light saturation of photosynthesis. In this case dark processes play the role of a governing slow stage, while energy migration and its transformation in reaction centers correspond to a fast stage.

2

Types of Dynamic Behavior of Biological Systems

The contemporary mathematical modeling has shown that the most informative model not "overloaded" with details typically involves two equations.

Just in the case when using the classification of variables into fast and slow ones it becomes possible to reduce the initial system to the following:

$$\begin{aligned}\frac{dx}{dt} &= P(x,y)\\ \frac{dy}{dt} &= Q(x,y)\end{aligned} \tag{2.1}$$

the qualitative methods of analyzing such systems are effectively used. Upon varying the state of the system in time, variables x and y change in accord with eq. (2.1) so that pairs of x and y values correspond to each state. In other words, by measuring variables x and y at consecutive moments of time $t_1, t_2, \ldots, t_n$ the state of the system is represented by appropriate pairs (x_1,y_1), (x_2,y_2), ..., and (x_n,y_n).

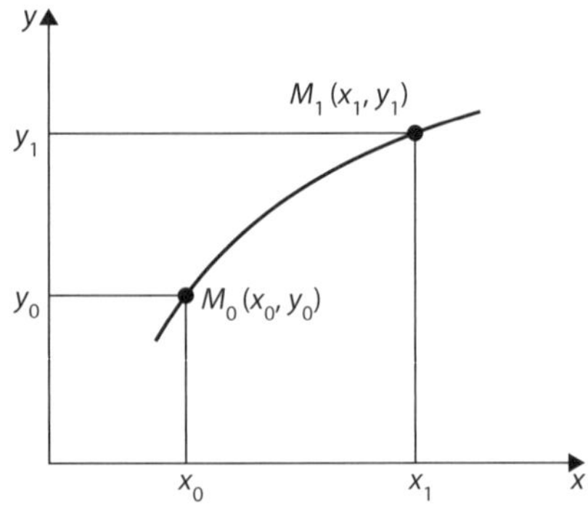

Figure 2.1 Region of phase trajectory on phase plane (*x*, *y*). *M*(*x*, *y*) is an image point.

Phase Plane Method. Let us analyze a plane with coordinate axes, variables *x* and *y* being plotted on them. Every point M on this plane with coordinates (*x*, *y*) corresponds to a certain state of the system. Such a plane is called a phase plane and point *M*(*x*, *y*) is called a representative point (Fig. 2.1).

Let at moment $t = t_0$ the coordinates of the representative point be $M_0(x_0\, y_0)$. At every subsequent moment t, the point *M* will move in conformity with equation system (2.1) and at every time moment it will occupy new position *M*(*x*,*y*) versus *x*(*t*) and *y*(*t*). The combination of all such points on the phase plane *x*, *y* is called a phase trajectory.

The character of phase trajectories illustrates typical qualitative features of the system behavior in time or, as thought, represents a "phase portrait" of the system. We will be interested in the phase portrait of the system nearby a stationary or singular point. In accordance with the definition of a stationary state, at a singular point simultaneously

$$P(x, y) = 0,$$
$$Q(x,y) = 0. \tag{2.2}$$

As a result, to find a singular point it is required to plot curves $P(x, y) = 0$ and $Q(x, y) = 0$ on the phase plane. The intersection of these curves will be the singular point and its coordinates will have stationary values $\overline{x}$ and $\overline{y}$. As an example let us analyze once again system (1.4) where variables *a* and *b* are designated as *x*=*a* and *y*=*b*.

The system can be rewritten as follows:

$$\frac{dx}{dt} = (k_1 - k_{+2}) + k_{-2}y - k_1 A = P(x, y),$$

$$\frac{dy}{dt} = k_{+2}x - (k_{-2} + k_3)y + k_3B = Q(x, y). \tag{2.3}$$

By equating the right-hand sides of (2.3) to zero according to (2.2) we obtain equations for curves $P(x, y)=0$ and $Q(x, y)=0$ for the localization of the singular point

$$P(x, y) = 0 \rightarrow y = \frac{1}{k_{-2}}[-(k_1 - k_{+2}) + k_1A] \tag{2.4}$$

and

$$Q(x, y) = 0 \rightarrow y = \frac{1}{k_{-2} + k_3}[k_{+2}x + k_3B].$$

Taking into account that $k_1 > k_{+2}$, equations (2.4) are written as

$$\begin{aligned} y &= -c_1x + c_2, \\ y &= c_3x + c_{4,} \end{aligned} \tag{2.5}$$

where c_1, c_2, c_3, and c_4 are positive constants readily determined from equations (2.4).

The plots of functions (2.5) are straight lines and their intersection point is stationary point $(\bar{x}, \bar{y})$ of the system (Fig. 2.2). In the case of second-order non-linear equations, function plots $P(x, y)=0$ and $Q(x, y)=0$ are not

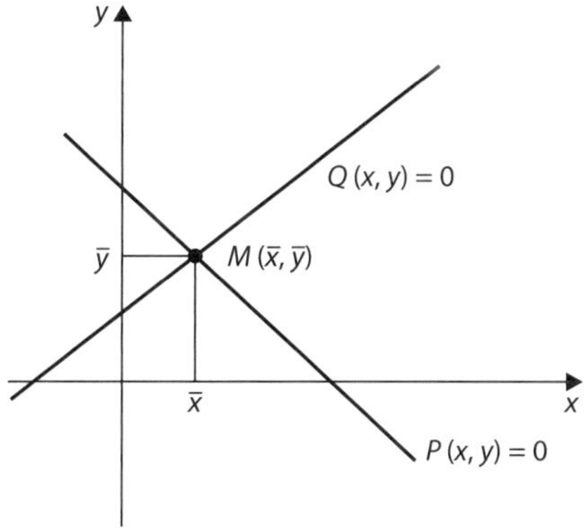

Figure 2.2 Localization of the special point of system (2.3).

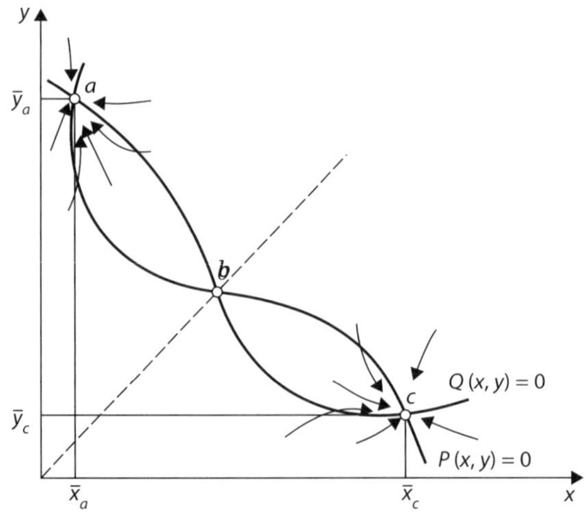

Figure 2.3 The case of three intersection points on a phase portrait. Phase portrait of a trigger system: *a* and *c* are stable singular points; *b* is an unstable singular point.

straight lines and can intersect with each other in several singular points, which corresponds to several stationary regimes (Fig. 2.3).

Types of Stability of Singular Points. Right now the task is to determine the type of stability of singular points from the expression of the right-hand sides of the initial equation system (2.1). To begin with it is clear that within the vicinity of a stable singular point, the representative point $M(x, y)$ will approach it by phase trajectories and, on the contrary, move away from it at unstable regime. So, in Fig. 2.3 special points *a* and *c* are stable, and *b* is unstable as seen from the direction of phase trajectories on the phase portrait.

Assume that the phase portrait has a special point $M(\overline{x}, \overline{y})$, the stability of which should be determined (Fig. 2.4). The same notions on the properties of stable states as those employed in studying the stability of stationary points with the equation containing one variable (see Chapter 1) will be used.

Let the system have a slight deviation from the stationary position $(\overline{x}, \overline{y})$ with some shifts (ξ, η) in variables x and y such as

$$x - \overline{x} = \xi, \quad y - \overline{y} = \eta. \tag{2.6}$$

By substituting these expressions in (2.1) the following system is obtained:

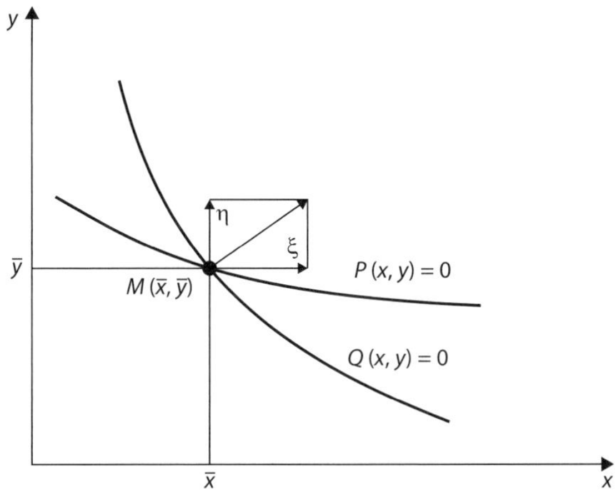

Figure 2.4 Determination of the stability of special point $M(\overline{x},\overline{y})$.

$$\frac{d\xi}{dt} = P(\overline{x}+\xi,\overline{y}+\eta),$$
$$\frac{d\eta}{dt} = Q(\overline{x}+\xi,\overline{y}+\eta). \tag{2.7}$$

Now it is necessary to expand the right-hand sides of (2.7) in the Taylor series nearby point $(\overline{x},\overline{y})$. As a result an equation for time-dependent values ξ,η is obtained:

$$\frac{d\xi}{dt} = a\xi + b\eta,$$
$$\frac{d\eta}{dt} = c\xi + d\eta, \tag{2.8}$$

where coefficients *a, b, c,* and *d* are the values of partial derivatives at point $(\overline{x},\overline{y})$:

$$a = P'_x(\overline{x},\overline{y}), \qquad b = P'_y(\overline{x},\overline{y}),$$
$$c = Q'_x(\overline{x},\overline{y}), \qquad d = Q'_y(\overline{x},\overline{y}) \tag{2.9}$$

It is evident that at this point it is required to take into account simultaneously the time behavior of deviations ξ and η of both variables *x* and *y*. However these deviations can change with time differently, which means that in the vicinity of the stationary point the phase trajectories will behave depending on the character of changes in both variables (ξ, η).

The general solution of system (2.8) for variables ξ and η is

$$\xi(t) = c_{11}e^{\lambda_1 t} + c_{12}e^{\lambda_2 t},$$
$$\eta(t) = c_{12}e^{\lambda_1 t} + c_{22}e^{\lambda_2 t}, \tag{2.10}$$

where powers λ_1 and λ_2 of exponents depend on the values and signs of both partial derivatives at point $(\overline{x}, \overline{y})$

$$\lambda_{1,2} = \frac{a+b}{2} \pm \sqrt{\frac{(a+d)^2}{4} + bc - ad} \tag{2.11}$$

If the radicand in (2.11) is negative, then $\lambda_{1,2}$ is a complex conjugate number. The behavior of variables ξ and η in (2.10) is compatible with the behavior of variables x and y nearby stationary point $(\overline{x}, \overline{y})$ and depends on exponents λ_1 and λ_2.

1. Exponents λ_1 and λ_2 are negative real numbers ($\lambda_{1,2} < 0$), therefore deviations of ξ and η from stationary position $\overline{x}, \overline{y}$ diminish with time. A singular point is called a **stable node** (fig. 2.5, *a*).
2. Exponents λ_1 and λ_2 are positive real numbers ($\lambda_{1,2} > 0$). In this case the stationary point is called an **unstable node** (fig. 2.5, *b*).

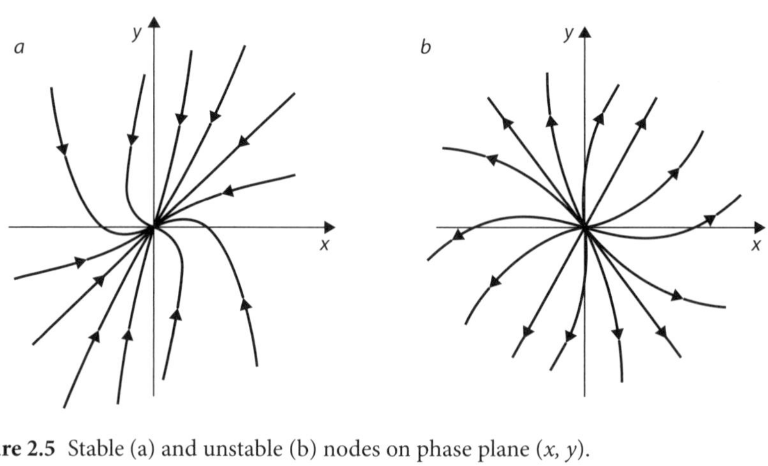

Figure 2.5 Stable (a) and unstable (b) nodes on phase plane (x, y).

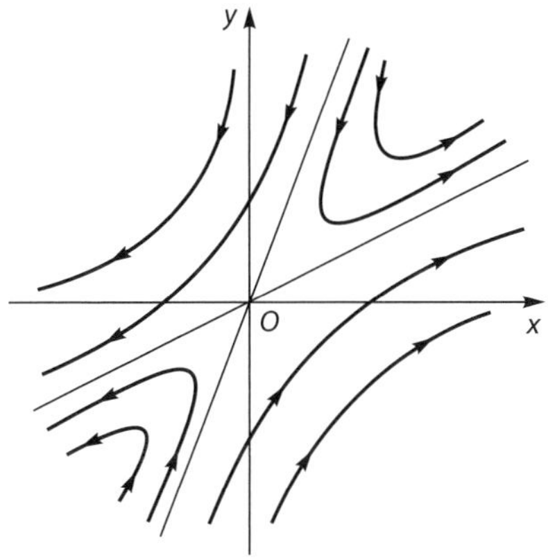

Figure 2.6 Singular point of the "saddle" type on phase plane (x, y).

3. Exponents are real numbers with opposite signs. So a singular point is called a **saddle** (fig. 2.6).

As seen, at any initial position on the plane (with the exception of a singular point itself and a separatrix) a representative point will move away from the stationary position.

If exponents are complex conjugate numbers, the time-dependent change in variables ξ and η will be oscillatory.

4. Real parts of $\lambda_{1,2}$ are negative (Re $\lambda_{1,2}<0$), deviations are damping. A singular point is a **stable focus** (Fig. 2.7, *a*).
5. Real parts are Re $\lambda_{1,2} >0$, the deviation amplitude increases with time. A singular point is an **unstable focus** (Fig. 2.7, *b*).
6. When Re $\lambda_{1,2} = 0$, the $\lambda_{1,2}$ values are purely imaginary numbers, and phase trajectories nearby the stationary point are ellipses that do not pass across the stationary point (2.8). This point is called a **center.**

Slight disturbances shift the system from one ellipsoidal trajectory to the other while the deviation amplitude changes and the "center" point is generally unstable.

The initial five types of equilibrium states are structurally stable; their character is not changed with slight changes in the right-hand sides of initial equations (2.1) and their first-order derivatives. In the case of the "center" in (2.11) we have

$$\text{Re}\,\lambda_{1,2} = \frac{a+d}{2} = 0,$$

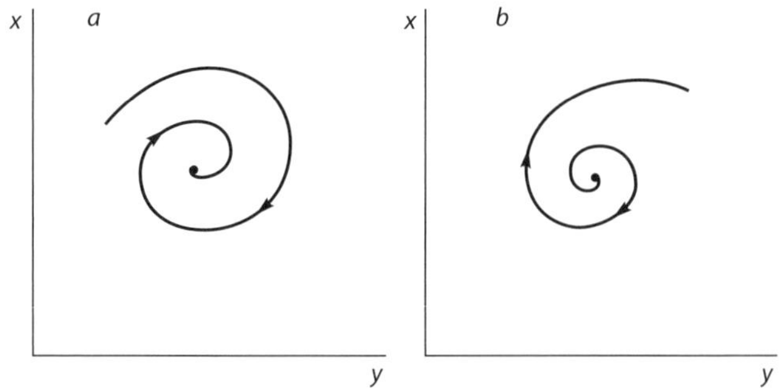

Figure 2.7 Stable (a) and unstable (b) "focuses" on phase plane (x, y).

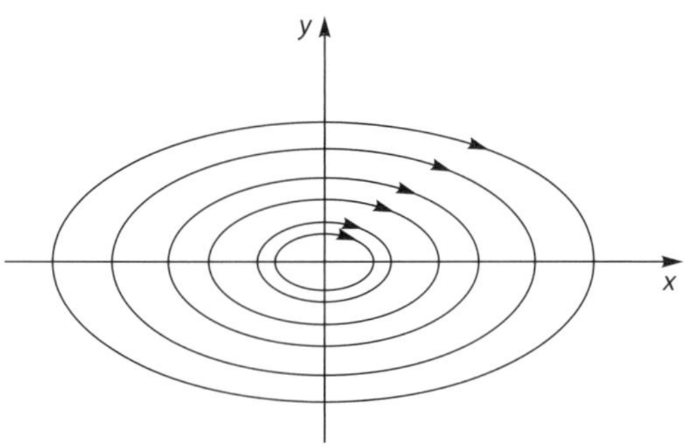

Figure 2.8 Singular point of the "center" type on phase plane (x, y).

which is valid only at $a = -d$, and also $bc < 0$, $|bc| > |ad|$. It is evident that slight changes in the right-hand sides of eq. (2.1) can easily infringe the ratios and thus alter the "center" type of the stability turning it into a stable or unstable focus. That is why the "center" singular point belongs to structurally unstable systems.

It is worth noting that stable biological systems should possess a definite "resource of structure stability", which permits them to retain basic dynamic properties at relatively small external disturbances.

Example 1. Consider an open system with a consecutive chain of transformations which proceeds according to the scheme

$$A \rightarrow x \rightarrow y \rightarrow B, \tag{2.12}$$

where A and B are external sources, and transformation of substance x into substance y occurs via the second-order reaction. If the inflow from the external source has constant rate v_0 and outflow y to the outside is described by the first-order equation, the following differential equations will characterize the open system (2.12)

$$\frac{dx}{dt} = v_0 - k_1 xy,$$
$$\frac{dy}{dt} = k_1 xy - k_2 y, \tag{2.13}$$

where k_1 and k_2 are rate constants. The coordinates of the stationary point can be readily determined from condition $\frac{dx}{dt} = 0, \frac{dy}{dt} = 0$:

$$v_0 - k_1 \overline{xy} = 0,$$
$$k_1 \overline{x}_2 \overline{y} - k_2 \overline{y} = 0. \tag{2.14}$$

They are
$$\overline{x} = k_2 / k_1, \ \ \overline{y} = v_0 / k_2. \tag{2.15}$$

Now it is necessary to determine the type of stability of this point. From equations (2.11) and (2.9) it follows

$$a = -\frac{k_0 k_1}{k_2}, \quad b = -k_2, \quad c = v_0, \quad d = 0,$$
$$\lambda_{1,2} = \frac{1}{2}\left(\frac{v_0 k_1}{k_2} \pm \sqrt{\left(\frac{v_0 k_1}{k_2}\right)^2 - 4 v_0 k_1}\right). \tag{2.16}$$

As seen, the stationary point is stable because $\operatorname{Re}\lambda_{1,2} = -\frac{v_0 k_1}{k_2} < 0.$ When $4 k_2^{\,2} > v_0 k_1$, the radicand is negative and the singular point is a stable focus, while at the opposite relation it is a stable node. So, dependent on the ratio of the rate constants, concentrations of carriers ($4 k^{\,2} > v_0 k_1$) demonstrate damping oscillations or monotonous approaching to the stationary state ($k_2^{\,2} < v_0 k_1$). The ratio of parameters $k_2^{\,2} = v_0 k_1$ acquires a bifurcational value, when the stability of a singular point of system (2.13) changes.

Example 2. Let us analyze the Volterra predator-prey model which illustrates alterations in the populations of prey (x) and predators (y) interacting with each other via the mechanism of "free collisions". This means that the population number of preys is proportional to the probability of their collision with predators, i.e. proportional to the product (xy). The same law regulates the increase in the population number of predators as a result of their collision with preys. In equation (2.13) this member agrees with the bimolecular reaction (of the kxy type). Moreover, natural mortality of predators occurs at the rate proportional to their population number, i.e. by the first-order reaction ($-ky$). The preys reproduce at a rate also proportional to their population number in conditions when the amount of their food is not limited.

Under such simplified suppositions, the appropriate equations are

$$\frac{dx}{dt} = \varepsilon_1 x - \gamma_1 xy, \qquad \frac{dy}{dt} = \gamma_2 xy - \varepsilon_2 y, \tag{2.17}$$

where ε_1, γ_1, γ_2, and ε_2 are corresponding rate constants or coefficients of proportionality.

Nonzero stationary values of coordinates of the stationary point of system (2.17) are as follows

$$\bar{x} = \frac{\varepsilon_2}{\gamma_1}, \quad \bar{y} = \frac{\varepsilon_1}{\gamma_2},$$

and the values of exponents $\lambda_{1,2}$ are purely imaginary

$$\lambda_{1,2} = \pm i\sqrt{\varepsilon_1 \varepsilon_2}.$$

This agrees with a singular "center" type point (see Fig. 2.8) which, as mentioned above, is unstable. This circumstance is a serious drawback of the Volterra model. To improve the model it is suggested, for example, to take into account self-restrictions in natural conditions of both population growth by the addition to the right-hand sides of equations the following second-order members ($-\gamma_3 x^2$), ($-\gamma_4 x^2$), that represent the effect of "tightness" and competition within the population:

$$\frac{dx}{dt} = \varepsilon_1 x - \gamma_1 xy - \gamma_3 x^2, \qquad \frac{dy}{dt} = \gamma_2 xy - \varepsilon_2 y - \gamma_4 y^2. \tag{2.18}$$

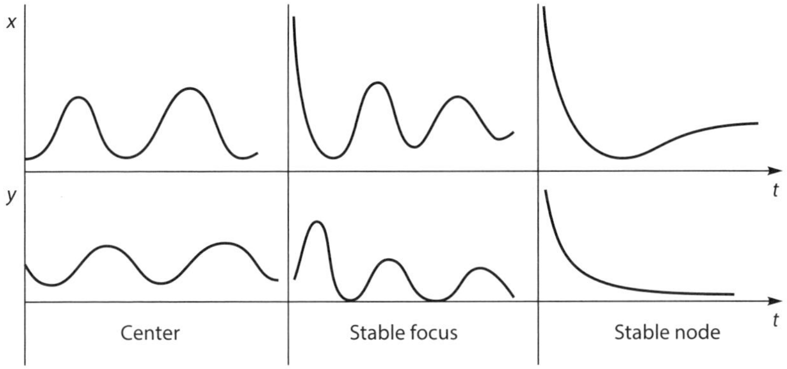

Figure 2.9 Specific changes in population numbers x and y at different types of special points.

Obviously at γ_3, γ_4 = 0 the system is reduced to a structurally unstable system with a *center* point. But at γ_3, $\gamma_4 \neq 0$, dependent on the relation between parameters, the system may have a *stable focus* point or, at high γ_3 and γ_4, values a *stable node.*

Figure 2.9 shows curves of time-dependent changes in the population number $x(t)$, $y(t)$ at different types of singular points in system (2.18). The appearance of even small non-linear members in right-hand sides of equations evokes qualitative changes in the phase portrait and transformation of the non-stable point of the center type into a stable one (focus or node). Parameters γ_3 and γ_4 can be regarded as actually governing ones that alter the type of stability of the stationary state. It is evident that in models of other systems it is also possible to find such governing parameters and their bifurcational relations which cause qualitative changes in the phase portrait of the system and the type of its stability.

Biological Triggers. Figure 2.10 shows the phase portrait of a system with three special points: *a* and *c* are stable nodes, and *b* is a saddle. The saddle is crossed by a dashed curve (separatrix) which divides the phase portrait of the system into domains of attraction of special points *a* (to the left of the separatrix) and *c* (to the right of the separatrix). If the system functions in a stable regime $a\,(\bar{x}_a, \bar{y}_a)$, it can be switched to regime $c(\bar{x}_c, \bar{y}_c)$, for example, by sharply increasing $\bar{x}_a$ (up to the $\bar{x}_c$ values in the range of domains of attraction of stable node *c*). After that the system switches by itself to regime *c* along one of the phase trajectories. This mode of switching the regimes is called a forced one, because some amount of the substance should be added to realize it. However, to switch the regime, one can also use the dependence of the phase portrait of the system on

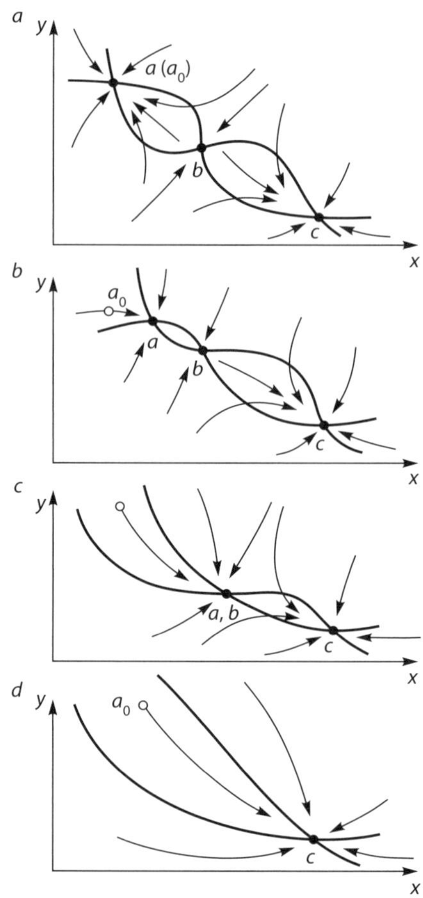

Figure 2.10 Process of parametric switching of the trigger system to the phase plane.

some governing parameter. Figure 2.10 demonstrates how the phase portrait of the system changes with alterations in this parameter. As seen, in the beginning points *a* and *b* merge with the formation of a complex point of the *saddle-node* type, and then only one stable point *c* remains on the plane to which the system would switch independently. Then it is possible to return to the initial values of the governing parameter and restore the phase portrait of the system, however it will be retained in the regime required.

The property of a trigger system to switch from one regime to the other makes it practical in designing models of biological processes when there is a number of plausible stationary states and transitions between them. Thus, during tissue differentiation, such switching of a cell takes place with

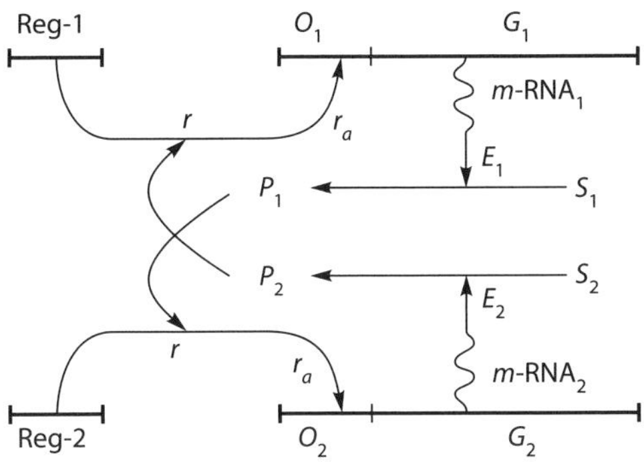

Figure 2.11 Scheme of mutual adaptation of two systems of enzymatic synthesis (Jacob and Monod model).

appropriate alteration in the yield of biosynthesis products. Figure 2.11 shows a scheme of protein synthesis regulation in prokaryotes. The product of the second system p_2 is a co-repressor of the first system. It associates into an active complex with the inactive repressor of the regulator gene of the first system, thereby blocking the operon of the structural gene in the first system. Similarly, the product of the first system p_2 is a co-repressor of the second system.

The analysis of the mathematical model has demonstrated that this system has trigger properties displayed when n (the order of the repression reaction relative to the concentration of products p_1 and p_2) is larger or equal to 2 ($n \geq 2$). This is the case when two or more product molecules are involved in co-repression.

Other examples of trigger systems will be examined in Chapter 3 devoted to the kinetics of enzyme processes.

Oscillatory Processes. Various biological systems demonstrate periodic processes: alterations in concentrations of intermediate products in glycolysis and photosynthesis, alterations in population numbers, and periodic biochemical reactions. The interest to oscillatory biological processes has been especially enhanced in studies of the "biological clock" based on the autooscillating system of intracellular biochemical reactions.

In all of the above cases, it is just internal dynamic properties of the system rather than some external disturbances that are the reason of fluctuation changes. Such systems are called autooscillatory. A closed curve on the phase plane corresponds to periodic motions. If this closed curve is isolated and the neighboring trajectories are approaching it from the outside and inside along a helix, such an isolated trajectory represents a limit cycle (Fig. 2.12). After small disturbances the system would return to

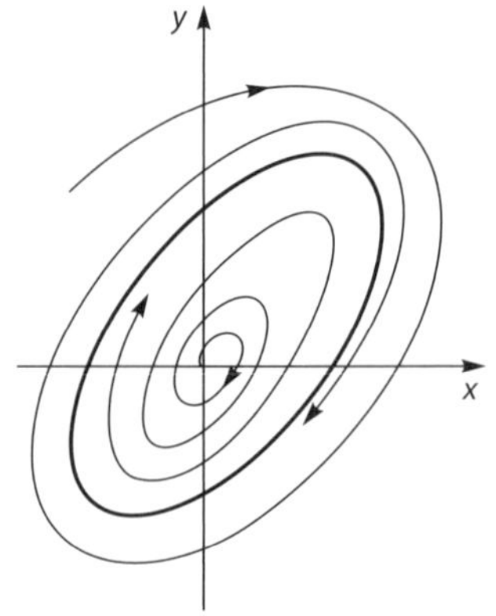

Figure 2.12 Stable limit cycle on phase plane *x*, *y* as the phase portrait of fluctuations in the system of glycolysis.

the trajectory of a stable limit cycle. This is what distinguishes it from the trajectory around the center point (Fig. 2.8), which is totally unstable. The periodicity and amplitude of movement along the limit cycle trajectory do not depend on the initial conditions.

Oscillations in Glycolysis. The glycolytic chain is a classic example of an oscillatory biochemical system. Introduction of sensitive spectrophotometry methods into biophysical investigations has allowed observing changes in concentrations of intermediate substances directly in the intact cell. The experiments on yeast cells demonstrated that the glycolysis processes go along with periodic changes in the concentrations of intermediate substances such as fluctoso-6-phosphate (F6P), fluctoso-1,6-diphosphate (FDP) and reduced NAD. These oscillations were observed most clearly on starving cells when the rate of glucose substrate uptake was low. Initial attempts of modeling the fluctuation processes in glycolysis were made without taking into account the bottleneck principle. The original model included 22 equations in accordance with the individual stages of glycolysis, and its analysis did not yield unambiguous results relative to the parameter values when oscillations appear. When the bottleneck principle was applied and the variables were separated into fast and slow ones, the simplified model included only the stage of glucose transformation into F6Ph and the stage of F6Ph transformation into FDP as a result of the action of enzyme phosphofructokinase (PFK).

$$\text{Gl} \rightarrow \text{F6P} \rightarrow \text{FDP} \rightarrow$$

3

Kinetics of Enzyme Processes

Enzyme processes play a key role in cell metabolism and, consequently, their kinetics should be of basic significance in the dynamics of cellular processes.

Michaelis-Menten Model. Let us analyze an elementary enzyme reaction in which one substrate (S) and one enzyme (E) are involved with one product (P) formed upon decay of the enzyme-substrate complex (ES):

$$E + S \underset{k_{-1}}{\overset{k_1}{\leftrightarrow}} (ES) \overset{k_2}{\rightarrow} E + P, \tag{3.1}$$

where k_1 and k_{-1} are constants of direct and inverse reactions of (ES) formation, and k_2 is the rate constant (v_p) of product P formation.

System (3.1.) is described by the following differential equations:

$$\begin{aligned}
\frac{dS}{dt} &= -k_1(S)(E) + k_{-1}(ES),\\
\frac{dE}{dt} &= -k_1(S)(E) + k_{-1}(ES) + k_2(ES),\\
\frac{d(ES)}{dt} &= k_1(S)(E) - k_{-1}(ES) - k_2(ES),\\
\frac{dP}{dt} &= k_2(ES) = v_P.
\end{aligned} \tag{3.2}$$

Since the total concentration of enzyme E_0 in the system remains constant, at any time moment the sum of concentrations of the free (E) and bound (ES) enzymes is

$$(E)+(ES)=E_0. \tag{3.3}$$

Let us see whether it is possible to apply the bottleneck principle or the principle of separation of the variables into fast and slow ones in system (3.1). In equations (3.2) it is easy to separate two groups of variables. Variables S and P show processes of the substrate consumption and corresponding formation of the reaction product. Variables E and ES describe the formation and decomposition of the enzyme-substrate complex. Let us analyze to what extent the groups of variables differ from each other by their characteristic times. The time value τ_E of the complex (ES) existence is determined by the fastest stage of its decomposition with the formation of product P, i.e. by the rate constant v_P of the product formation. The same fast stage determines regeneration and the time of the existence of the free enzyme in the system. In other words, characteristic lifetime τ_E of variables E and ES is

$$\tau_E = 1/k_2. \tag{3.4}$$

Constant k_2 corresponds to the number of catalytic cycles, i.e. the number of (ES) decompositions and P formations per unit of time. Here the τ_E time is the time of the enzyme turnover and constant k_2 is called the number of enzyme turnovers. Characteristic time τ_S of S depletion in the system and the corresponding generation of P, may be dependent on the rate of the product formation. It is

$$\tau_E = S/v_P. \tag{3.5}$$

The maximal rate of the product formation will be reached when the entire enzyme E_0 is in the bound state, i.e.

$$v_P^{\max} = k_2 E_o$$

In these conditions, time τ_S of S depletion will be the minimal

$$\tau_S^{\min} = \frac{S}{k_2 E_0}. \quad (3.6)$$

Now we can compare the values τ_S and τ_E. It is evident that

$$\tau_S^{\min} = \frac{S}{k_2 E_0}, \quad (3.7)$$

Typically the substrate S concentration exceeds many times that of the product

$$[E_0] \sim 10^{-6} M,\ [S] \sim 10^{-2} M.$$

Hence it follows that $E_0 / S - \varepsilon \sim 10^{-4} \ll 1$, i.e.

$$\tau_S^{\min} = \tau_E \cdot \frac{1}{\varepsilon} \sim \cdot \tau_E 10^4 .$$

Normally the order of values $\tau_E \sim 10^{-2}$s and $\tau_S \sim 10^{2}$s shows that

$$\tau_S \gg \tau_E. \quad (3.8)$$

However, in accord with (1.21) this means that (E) and (ES) are fast variables. Their changes are so prompt, that all the time they remain near their stationary values $\bar{E}$, $\bar{E}\bar{S}$ Consequently, they can be described by algebraic equations obtained by equating to zero the right-hand sides of the second and third equations in model (3.2):

$$\frac{d(E)}{dt} = \frac{d(ES)}{dt} = 0. \quad (3.9)$$

The obtained result can be represented in two time scales τ_E and τ_S using the following fig. 3.1.

A small cart (E) carries cargo from one place (S) to another (P). It is rapidly loaded and unloaded (τ_E is low) and returns to be loaded again. But the general process of carrying the whole cargo from one place to another (S depletion and P emergence) proceeds at a much slower rate than each

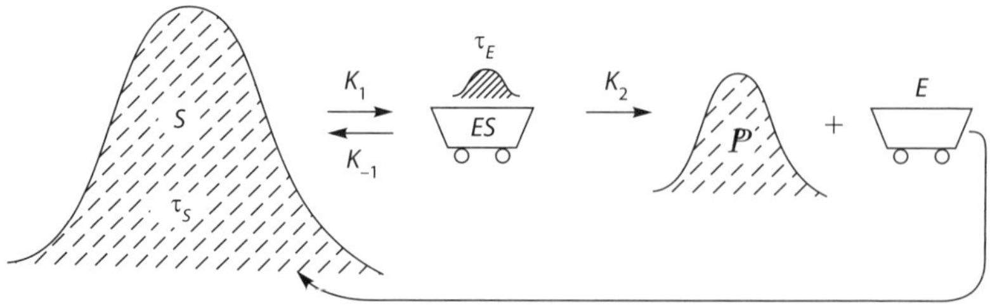

Figure 3.1 Two time scales of enzyme process (see explanations in the text).

individual working cycle of the cart ($\tau_E >> \tau_S$). Namely when conditions of (3.8) and (3.9) are valid it is possible to obtain the known Michaelis-Menten equation of the dependence of the stationary rate of enzyme reaction on the substrate concentration

$$\bar{v}_P = \frac{k_2 E_0 S}{K_M + S}, \qquad (3.10)$$

where $K_M = \frac{k_{-1} + k_{+2}}{k_1}$ is the Michaelis constant. From (3.10) it is seen that with an increase in the substrate concentration when $S \to \infty$, $\bar{v} \to \bar{v}_P^{\max} = k_2 E_0$, i.e. the reaction rate is saturated (Fig. 3.2). Equation (3.10) shows also that at K_M = S the reaction rate is $v_P^{\max}/2$. As a result, the Michaelis constant K_M is numerically equal to the substrate concentration, when half of the enzyme molecules are bound in complexes with the substrate and the stationary rate reaches half of its maximum value (see Fig. 3.2). For two different enzymes, the substrate concentration required to attain the same rate v_P is higher the lower is the activity of the particular substrate, i.e. the higher amount of it is required to form the particular amount of the enzyme-substrate complexes. Apparently, in (3.1) the ability to bind the substrate is assigned by constant k_1, its increase results in the enhancement of the enzyme activity and diminishing the K_M value.

Regulation of Enzyme Reactions. Under real conditions of cellular metabolism, the concentration of substrate used in enzyme reactions changes not only as a result of the reaction itself, but also due to its inflow into the reaction volume. Simultaneously the product outflows from the reaction area to other areas where it is consumed in further metabolic transformations. In other words, within the cell, every individual reaction, the same as their general totality, represents an open system with self-regulation mechanisms. One of the most powerful ways of regulation of enzyme process is to change

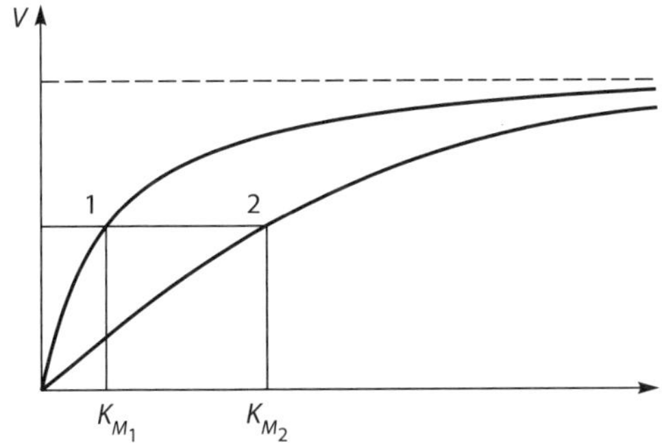

Figure 3.2 Dependence of the rate (v) on the substrate concentration (S). Curves 1 and 2 correspond to two enzymes with different Michaelis constants $K_{M2} < K_{M1}$.

the activity of the enzyme by different inhibitors. As known, there are competitive inhibitors which occupy the substrate positions in the active center of the enzyme with the formation of an inactive complex. Non-competitive, or allosteric inhibition, may also take place, when the inhibitor has no affinity to the substrate and is associated not with the active center of the enzyme but with particular sites of the protein globule, leading to the deformation of the enzyme. Regulatory effects can proceed by the feedback principle as well, when at high substrate or product concentrations the reaction is suppressed. Along with the inhibitors, there are also activators or substances enhancing the enzyme activity. Besides the reaction product per se can have an activating effect (activation by the product).

The above peculiarities and the nonlinear kinetics underlie the basis of mechanisms regulating enzyme processes and, as a result, control the dynamic behavior of complex combination of metabolic transformations in the cell. We will see that such regulation of even relatively simple enzyme processes brings about nontrivial types of dynamic behavior.

Open Enzyme System with Substrate Suppression. Such a system involves irreversible reaction of substrate inflow ($S_0 \xrightarrow{k} S$), product outflow from the reaction area and formation of the inactive complex

$$(ES) + S \underset{k_{-4}}{\overset{k_4}{\leftrightarrow}} (ES^2)$$

$$\begin{aligned} &S_0 \rightarrow S + E \underset{k_{-1}}{\overset{k_1}{\leftrightarrow}} ES \overset{k_2}{\rightarrow} E + P \overset{k_4}{\rightarrow}, \\ &ES + S \underset{k_{-3}}{\overset{k_3}{\leftrightarrow}} ES^2. \end{aligned} \tag{3.11}$$

The rate v_p of product formation in scheme (3.11) differs from that of the stationary Michaelis-Menten equations (3.10)

$$v_P = \frac{k_2 E_0 S}{K_M + S + \frac{S^2}{k_S}} \tag{3.12}$$

where $k_S = k_3/k_{-3}$. It should be noted that all members with S to the second power (S^2/k) specify strong nonlinearity of the system and the appearance of cubic members (S^3) in equations. At small S ($S <<\ 1$), when the S_2/k_S value in the denominator may be neglected, equations (3.10) and (3.12) are compatible. But with the growth of S an inactive complex ES^2 is formed and the rate of product formation $v_P \to 0$ at $S \to \infty$. The plot of the $v_P(S)$ dependence is a bell-shaped curve with a maximum. The rate of changing the substrate concentration S is summed of rate v_S of its inflow from the outside v_{infl}

$$v_{\text{infl}} = k\,(S_0 - S) = \alpha - kS \tag{3.13}$$

and its outflow in the process of the enzyme reaction ($v_{\text{outfl}} = v_P$):

$$v_S = v_{\text{infl}} - v_P.$$

The plot of dependence $v_{\text{infl}} = \alpha - kS$ is a straight line. At the stationary state of an open system (3.11) when the substrate concentration remains constant, the following equation should be valid:

$$\bar{v}_{\text{infl}} - v_P = 0.$$

Stationary points $\bar{S}$ at $\bar{v}_S = 0$ are intersection points of curves $v_P(S)$ and $v_{\text{infl}}(S)$. In fig. 3.3 the family of parallel curves corresponds to different α values. By changing α or the external concentration of substrate S_0 one can see that the system may possess from one to three stationary states.

By changing the value of the governing parameter α it is possible to change the number of stationary points. In the range $\alpha_2 < \alpha < \alpha_4$, three various stationary states correspond to every α value whereas the values α_2 and α_4 when the number of stationary points changes are bifurcational.

Stationary points vary in their stability. From the three stationary points $\bar{S}'$ $\bar{S}_3''$ and $\bar{S}_3'''$, point $\bar{S}_3''$ at $\alpha = \alpha_3$ is unstable, while $\bar{S}_3'$ and $\bar{S}_3'''$ are stable. This can be confirmed by the analysis of its stability (see Fig. 3.3). Indeed, let assume that due to deviation $\Delta S < 0$ from point $\bar{S}_3'$ the S value becomes lower than stationary value $\bar{S}_3'$. In the range $S < \bar{S}_3'$ the rate of the

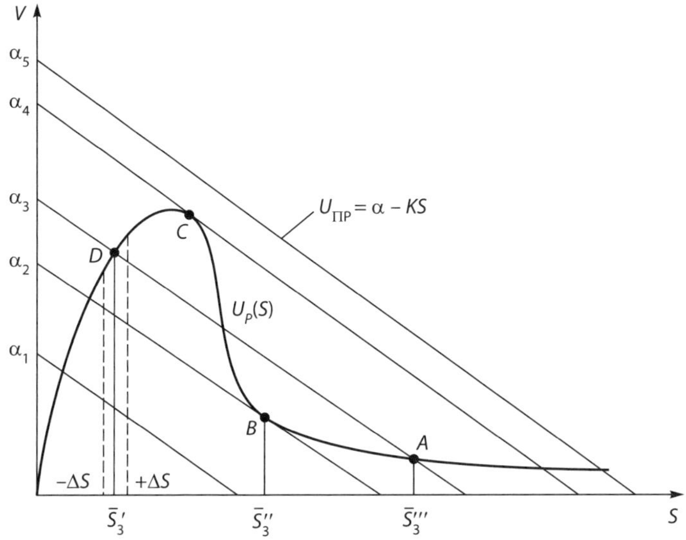

Figure 3.3 Dependence of the number of stationary states versus parameter α and their stability in a system with substrate suppression and reversible substrate inflow.

substrate inflow is higher than that of its outflow ($v_{infl} > v_p$) and, as a result, variable S will grow approaching the $\bar{S}_3'$ value. If the deviation from the stationary point is $\Delta S > 0$, i.e. $S > \bar{S}_3'$, then in this range $v_{infl} < v_p$ and S will decrease coming back to stationary value $\bar{S}_3'$. Thus, at any deviation from stationary state $\bar{S}_3'$, the system would return to it, and consequently state $\bar{S}_3'$ is stable. Similar analysis demonstrates that point $\bar{S}_3''$ is stable whereas point $\bar{S}_3'''$ is unstable. The stationary stable points are located on curve $v_p(S)$ at regions AB and CD, whereas the unstable points are at region BC. It is possible to plot the $\bar{S}_3(\alpha)$ dependence of stationary states versus governing parameter α. Figure 3.4 shows that with a change in α the system may switch from one stable regime to the other.

Let us assume that in the initial state the system is at point A at the upper branch of curve $\bar{S}_3(\alpha)$. When the inflow rate decreases and α diminishes, the system will "shift" to the left along the upper stable branch AB of stationary states. When the bifurcation value of α_2 is reached, the system would leave the unstable point B and would make jump-like transition $B \rightarrow D$ to the lower branch of stable stationary states. By further increasing α_2 to α_1 the system can be transmitted along stable branch DC to the bifurcation point C. After reaching this point the system would spasmodically return to the initial state on upper branch AB.

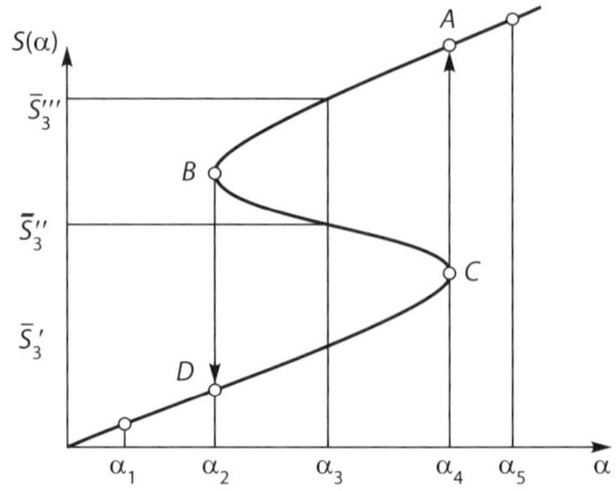

Figure 3.4 Curve of stationary states $\bar{S}(\alpha)$ in a system with substrate suppression and reversible substrate inflow.

So, the system can function in one of the two stable stationary states and therefore it has trigger properties. The direction of transitions and realization of one or the other stable states depend on whether the αparameter decreases or increases. The system can perform transitions from one state to the other in different ways depending on the preceding increase ($DC \rightarrow AB$) or decrease ($AB \rightarrow DC$) of the α parameter or on the system history. This property is called hysteresis. The periodical movement along the hysteresis cycle $A \rightarrow B \rightarrow D \rightarrow C \rightarrow A$ has the oscillatory nature.

Strong nonlinearity caused by mechanisms of substrate suppression, as well as the existence of fast and slow variables provoke the emergence in the system of multiple stationary states and trigger and hysteresis features.

Oscillations in Enzyme Systems are produced by analogous reasons. Since the substrate concentration in enzyme systems is as a rule far higher than the enzyme concentration, this promotes the existence of two drastically varying time scales for the substrate and the enzyme-containing complexes. A necessary condition for undamped oscillations is an open nature of the system as closed enzymatic systems may have only damped oscillations.

The most prevailing reason for nonlinearity of biochemical systems is associated both with the existence of reversible links in the reaction chain when the product suppresses or activates the reaction and with the substrate suppression. These factors strongly destabilize the systems promoting the appearance of an unstable point. In the vicinity of the latter a limit cycle can be formed. A stable autooscillatory regime is conditioned by the unstable state of the system.

Let us analyze a reaction scheme with the substrate and product suppression

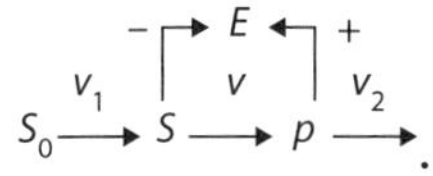

(3.14)

The change in the concentration of slow variables of the product and substrate is described by equations

$$\frac{dS}{dt} = v_1 - v(S,P),$$
$$\frac{dP}{dt} = v(S,P) - v_2. \quad (3.15)$$

Note that S and P should be represented as dimensionless values ($S/\bar{S}$, $P/\bar{P}$). But for the sake of simplicity, we avoid the use of dimensionless variables. In this system, suppression proceeds as a result of non-competitive inhibition with the product and substrate in line with the scheme

$$S + ES \rightleftarrows ESS,$$

$$P + ES \rightleftarrows ESP,$$

$$P + E \rightleftarrows EP,$$

$$P + ESS \rightleftarrows ESSP.$$

When the reaction of the substrate inflow is reversible, the product outflow will be linear. The dependence $v(S)$ has a bell-like shape with the maximum.

It follows from the analysis that the number of stationary states of the system and their stability depend on the depth of the product suppression. At weak suppression by the product, the relative concentration of the substrate is a fast variable as compared to the concentration of the product. As this takes place, the only unstable stationary state is located at the unstable part of the characteristic $v(S)$. Self-sustained oscillations are generated around the unstable state on the phase plane S, P (Fig. 3.5) in the system during its transfer along the cycle $C \to A \to D \to B \to C$. Points A and B are positioned at the boundaries of stable (AC and DB) and unstable (AB) branches of quasistationary curve ($\dot{S}=0$). The movement along the CA branch proceeds towards point $A(C \to A)$ with the product

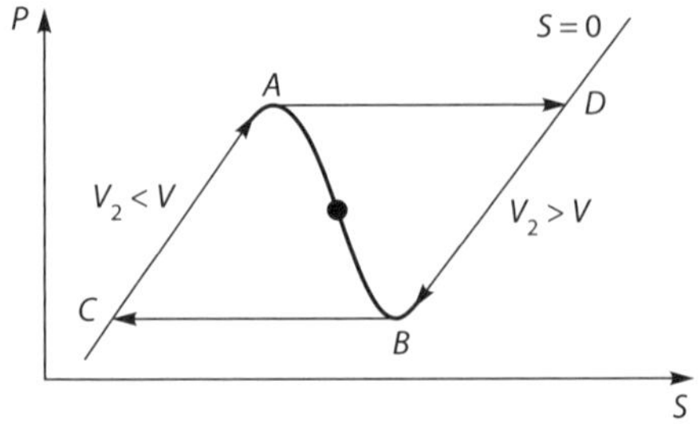

Figure 3.5 Limit cycle on phase plane of system (3.14) at low depth of product suppression.

accumulation because in the *CA* range, rate v_2 of the product outflow is lower than the rate of its formation. At the *A* critical point when $v=v_2$, the system looses its stability and abruptly comes to point *D* of branch *DB* where the outflow rate v_2 is higher than the reaction rate. Due to this, the product concentration begins diminishing, while rate v grows. Having reached point $B(v=v_2)$, the system loses its stability once again and "falls" to a fast movement towards the initial point *C*. Then the cycle recurs and the system performs self-sustained oscillations.

General conditions of oscillations in enzyme reactions suggest the presence of a large number of intermediate stages in an open chain. It has been found that this is realized in polyenzyme ring systems in which the final product determines the rate of the initial stage.

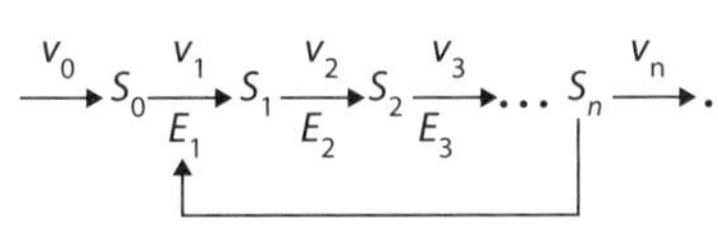

(3.16)

The feedback arrow shows the suppressive effect of the final product on the activity of key enzyme E_1.

The nonlinear kinetics, feedback regulation, hierarchy of characteristic times, and changes in the variables specify different types of biological dynamics. Different biological processes are often characterized by similar dynamical properties, typologically similar phase portraits and their dependencies on controlling parameters. Based on this, it is possible to classify biochemical systems by the types of their dynamical parameters. For example, some typical mathematical models are analogous in their formal structure to scheme (3.14) that may be used for the comparative studies to determine parametric characteristics of their dynamical behavior.

4

Distributed Biological Systems. Chaotic Processes

Let us analyze models in which variables change not only in time but in space as well. In contrast to point models, such models are called distributed (spatially). In distributed systems, chemical transformations of substances can occur parallel with the diffusion of individual substances from elementary volumes with high concentrations to those with low concentrations. So, neighboring volumes are connected because of the transition processes. In biological systems (such as active membranes, tissues, and communities of organisms), there are also distributed sources of energy. Part of this energy is dissipated in elementary volumes of the system. Such systems are recognized as active distributed systems.

An example of biological process occurring in a distributed system is structure formation in morphogenesis. It occurs spontaneously based on the information, contained in the fertilized oocyte, in the initially spatially homogeneous medium rather than because of external stimuli. In this case, we have in mind the emergence of stationary three-dimensional non-homogeneous structures in the active distributed system. Another example is propagation of excitation waves in nerve and muscle tissues.

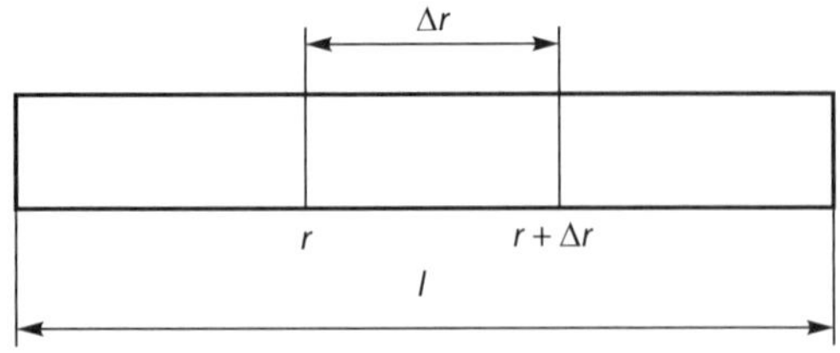

Figure 4.1 Distributed system with one variable x involved in the chemical process and diffused along a narrow tube.

Equation for Distributed System. The analysis of simple systems has demonstrated that different behavior of active distributed systems can be described with nonlinear differential equations in partial derivatives which take into account chemical reactions and diffusion of reagents.

Let us consider a simple case of one variable x that is involved in the chemical process and simultaneously is diffused along a narrow tube (Fig. 4.1). The diffuse flow of the substance (i.e. the mass passing in a time unit across a unit of area perpendicular to the diffusion direction) is proportional to the concentration gradient of this substance with the opposite sign

$$I = -D\frac{\partial c(r,t)}{\partial r}, \tag{4.1}$$

where D is the diffusion coefficient.

It can be demonstrated that a time-dependent change in the substance concentration, caused by diffusion processes in the tube elementary volume between points r and $r+\Delta r$, depends on the difference in flows I at points r and $r+\Delta r$ and in the limit at $\Delta r \rightarrow 0$ is

$$\frac{\partial c}{\partial t} = -\frac{\partial I}{\partial r} = \frac{\partial}{\partial r}\left(D\frac{\partial c(r,t)}{\partial r}\right).$$

If diffusion coefficient D is constant, then the equation for diffusion is as follows

$$\frac{\partial c}{\partial t} = D\frac{\partial^2 c(r,t)}{\partial r^2}. \tag{4.2}$$

The above equation describes the time-dependent change in the substance concentration when only diffusion proceeds in the system. However in addition to diffusion, chemical reactions also take place with "point" members $f(c)$ corresponding to them in a simple case. The general equation for the "c" change caused by chemical reactions and diffusion is the following

$$\frac{\partial c}{\partial t} = f(c) + D\frac{\partial^2 c}{\partial r^2}. \tag{4.3}$$

If the system contains several substances $c_1, c_2, \ldots. c_n$, then in place of (4.3) one should write

$$\frac{\partial c_i}{\partial t} = f_i(c_1, c_2, \ldots c_n) + D_i\frac{\partial^2 c_i}{\partial r^2}, \quad i = 1, 2, \ldots, n. \tag{4.4}$$

Analysis of Models of Distributed Systems is a different problem which we discuss here only in the most general terms. To solve systems of typical differential equations (point models), it was required to take the initial values of variables at the zero time $t = t_0$. In the case of distributed systems, edge or boundary conditions should be assigned at the boundary of the area where the studied process takes place. The boundary conditions depend on the way the substance concentration at the boundary is changed. For example, at the end of the tube there may be either constant concentration of the substance in the vessel the tube has contacts with, or the butt-ends of the tube are impermeable for the diffusion flow.

It should be noted that distributed system (4.4) can be reduced to the point system

$$\frac{\partial c_i}{\partial t} = f(c_1, c_2, \ldots c_n); \quad i = 1, 2, \ldots, n.$$

if all diffusion coefficients $D_i = 0$ or if, to the contrary, they are very large, so that all initial reagents and products have time to be mixed completely in the whole reaction volume during the chemical reaction. Stationary points should be found from the condition that time derivatives $\frac{\partial c_i}{\partial t} = 0$:

$$D_i\frac{\partial^2 c(r,t)}{\partial r^2} + f_i(c_1, c_2, \ldots, c_n) = 0, \tag{4.5}$$

where $\bar{c}_i(r)$ can be determined from stationary values. After that some disturbance $\Delta c_i(r)$ is produced and its behavior in time is analyzed. If with time at $t \rightarrow \infty$ the inserted slight disturbance $\Delta c_i(r)$ is not enhanced in the system, the initial stationary point $\bar{c}_i(r)$ would remain stable. The behavior of the initial deviation depends on the properties of functions $f_i(c_1,c_2,\ldots,c_n)$ and diffusion coefficients. In particular, in a one-dimensional problem, at $f_c'(\bar{c})<0$ the initial deviation in time damps at $t\rightarrow\infty$.

Basic Models. It is impossible to describe complex behavior of variables, e.g. the oscillatory state of the system, by using one equation. The principal results in studying the properties of distributed systems were obtained on the so-called "basic models" with two variables (cf. (2.1)):

$$\frac{\partial x}{\partial t} = P(x,y) + D_x \frac{\partial^2 x}{\partial r^2},$$

$$\frac{\partial y}{\partial t} = Q(x,y) + D_y \frac{\partial^2 y}{\partial r^2}. \tag{4.6}$$

It has been found that based on a simple model like (4.6) it is possible to obtain a qualitative description of the processes of spontaneous generation of waves and structures in distributed systems, i.e. the processes of self-organization. They occur when instabilities appear in the system leading to the loss of the initial distribution of substances in time and space. Instead, a new type of substance distribution in time and space is formed, that is the system is self-organized. For instance, the loss of stability of the stationary homogeneous distribution of substances in space in a chemical reaction may lead to the generation of autowaves (periodic self-sustaining waves of chemical activity) in the system.

Depending on the type of functions $f_i(c_1,c_2,\ldots,c_n)$ and diffusion coefficients D_i, the following nontrivial types of behavior of variables and types of self-organization can arise in the system.

1. Expansion of disturbances as a traveling pulse.
2. Standing waves.
3. Synchronous autooscillations of different elements in the entire space.
4. Quasistochastic waves that are produced at random disturbance of the phase difference of autooscillation at two space loci.

5. Stationary nonhomogeneous distributions of variables in space, i.e. dissipative structures.
6. Wave generation by autonomous sources of pulses. Local short-term fluctuations of variables can play the role of wave sources.

A general condition for processes of self-organization is the development of instability in the initial distributed system. Specifically, the development of instability of the saddle type results in the appearance of dissipative structures, and the formation of an unstable node may cause generation of finite amplitude traveling waves or standing waves. The dissipative structure formed as a result of instability of the distributed system is preserved by the continuous inflow of the energy and substance and can exist only in open systems. This is what distinguishes it from common equilibrium structures. The formation of such dissipative structures underlies tissue differentiation during morphogenesis. The abrupt transitions between dissipative structures of different shapes, induced at the increase in the length of the reaction vessel, represent the principal characteristic of cell division processes.

The Brusselator is the most studied system that demonstrates different dynamic behavior in time and space at diverse values of its parameters. The brusselator model allows clarifying conditions for self-organization in biological and chemical systems, and to this effect the model is basic. It should be underlined that the brusselator is characterized by simple cubic nonlinearity specified by the reaction

$$2x + y \rightleftarrows 3x.$$

An example of such reaction may be an enzyme process in which the enzyme has at least three catalytic centers. Cubic nonlinearity is a critical condition for the formation of dissipative structures.

The brusselator represents the following scheme of hypothetical chemical reactions:

$$\begin{aligned} &A \rightarrow x, \quad 2x + y \rightarrow 3x, \\ &B + x \rightarrow y + C, \quad x \rightarrow R, \end{aligned} \tag{4.7}$$

where A and B are initially agreed substances, distributed uniformly within the tube, and their supply is large; substances R and C precipitate. Variables x and y diffuse along the tube and are also involved in chemical processes.

The brusselator model reads like this

$$\frac{\partial x}{\partial t} = A + x^2 y - (B+1)x + D_x \frac{\partial^2 x}{\partial r^2},$$
$$\frac{\partial y}{\partial t} = Bx - x^2 y + D_y \frac{\partial^2 y}{\partial r^2}. \quad (4.8)$$

Let us describe the results of studying the types of the above model behavior as dependent on the ratio of parameters (A, B, D_x, D_y). The point model (D_x=D_y=0) has a stationary point

$$\bar{x} = A, \bar{y} = B / A. \quad (4.9)$$

At $B<1+A^2$, this point represents a stable focus, while at $B>1+A^2$ it is an unstable focus with a limit cycle formed around it in the point system.

In distributed system (4.8), there may appear instability of a saddle type that results in perturbations and three-dimensionally unstable stationary regimes in a homogeneous system. At certain sizes of the tube and wave lengths, determining the type of non-homogeneities or space irregularity, time-independent periodic dissipative structures may originate in the system. This becomes possible if coefficients D_x and D_y are essentially different while parameters A and B are too far from their bifurcation values. Moreover, autowave processes like standing or traveling waves may also occur in the brusselator. Continuous changing of the parameters is the reason for different dissipative structures to replace each other. Therefore, the brusselator is a typical trigger with many stable states, i.e. types of dissipative structures.

Relying on the basic model, the process of cell division can be associated with parametrically assigned formation of a new dissipative structure, the recurrence from which is impossible because of hysteresis.

It is believed that the formation of dissipative structures depends on the dynamic reading of the parametrically defined information. Wave processes can be observed also in direct experiments as demonstrated by the Belousov-Zhabotinsky redox reaction in the presence of bromo-malonic acid and cerium or manganese ions as catalysts. These processes are described by equations of the basic type (4.8) as well.

Below (Chapter 5) we will analyze models of formation of dissipative structures in ecological systems.

Chaotic Processes in Determined Systems. In the systems considered above, the determined character of their behavior means that small changes in the initial conditions lead to small alterations in the final result. In essence, this is also valid for the behavior of the systems at bifurcation points. In fact insignificant perturbations "push" the system to one of the regimes *a priori* possible. In such cases (trigger switching and stable limit cycle near the unstable focus), the corresponding phase trajectories have a quite definite shape, while only the transition of the system to one of the regimes is accidental.

However it has become clear that a rather common feature of many nonlinear determined systems is their chaotic behavior in time. It has appeared that random behavior of many biological, meteorological and economic nonlinear models with the number of degrees of freedom exceeding two is observed at the specific critical values of their internal parameters. This means that in sufficiently long time periods, the behavior of the systems becomes unpredictable so that the systems themselves are in the state of chaos and irregularity. It should be underlined that in this case chaos appears as a result of internal dynamics of the system and not as a consequence of external sources or external random actions.

At present it becomes clear that the chaotic behavior is inherent to a wide range of determined systems and plays a great role in self-organizational processes in nature.

Models of Chaotic Systems. We will describe briefly some models of determined systems with chaotic behavior and discuss their potential significance in biology. Note that as far back as at the end of the 19-th century Poincare discovered that some mechanical systems governed by equations of classical mechanics demonstrate chaotic behavior.

For example, a mechanical pendulum periodically forced by external power $F_0 \cos\omega t$ displays random dependence on the angle θ versus time (Fig. 4.2) if amplitude F_0 of the driving force exceeds some threshold value F_c.

In other experiments, the motion in layers of the liquid in a capsulated vessel heated from below was examined. At a large difference in temperatures ΔT between the top hot and lower cold layers, the stationary convective motion disappears and transition to random motion, i.e. Bernard instability, is observed (Fig. 4.3). In the Belousov-Zhabotinsky reaction, the stationary three-dimensional distribution of stained reagents (cerium ions) is broken at definite flow rates of the reaction mixture through the reactor, and a chaotic regime is formed in the system. All these processes are described by systems of autonomous nonlinear differential equations of the first order. The analytical investigation allowed determining qualitative characteristics of random motion that is established upon changing the external control parameter (the amplitude of the driving force F_0, the difference in temperatures ΔT). Here, the following questions arise: Are there general regularities for the transition of determined systems to chaotic states? Is it possible to predict the probability of randomness proceeding from the type of differential equations of a determined model? What is the role of chaos in the behavior and evolution of determined systems?

Apparently, the first strict determined model with chaotic behavior was the system of equations proposed by Lorenz (1963) for weather forecasting in meteorology. The model is based on the ideas that air flows in the atmosphere are connected with the difference in temperatures of its various layers. This approach may be used for describing the behavior of the liquid heated from below in the Bernard experiment (Fig. 4.3).

The Lorenz model reads like this

$$\dot{x} = -\sigma x + \sigma y, \dot{y} = rx - y - xz, \dot{z} = xy - bz \quad (4.10)$$

$$b\dot{y} = rx - y - xz$$

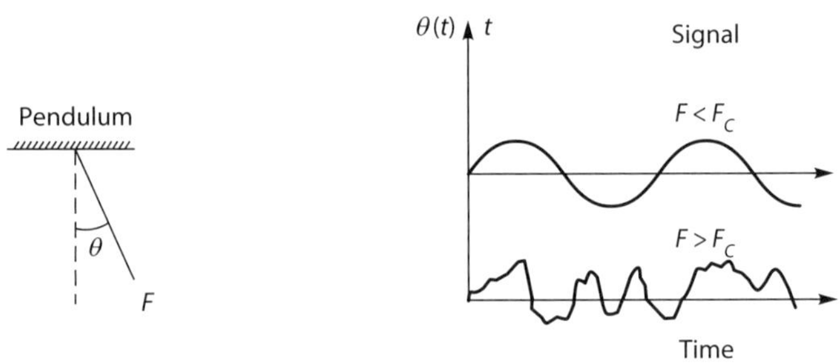

Figure 4.2 A periodically forced pendulum. Dependence of the driving force F and angle θ versus time (according to Schuster H.G., 1984).

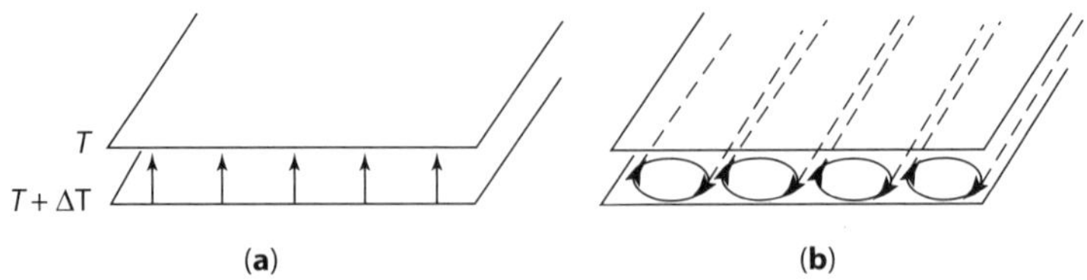

Figure 4.3 Bernard instability (according to Schuster H.G., 1984).
(a) Heat flows; (b) convective waves generated in the liquid when temperature gradient T is higher than the critical value.

where σ and b are dimensionless constants, and r is the control parameter proportional to the difference in temperatures. Variable x is proportional to the velocity of the circulating liquid; y is the difference in temperatures between the upstream and the downstream; and z is proportional to the deviation of the vertical temperature profile from the equilibrium value.

In this model, variables can behave chaotically at an increase in the difference in temperatures ΔT, when the value of the control parameter r_c is higher than the critical value r_c $(r>r_c)$. Figure 4.4 shows how variable y behaves in time. At $r<r_c$ dependence $y(t)$ represents a damping periodical motion. However when critical threshold $r>r_c$ is surpassed, irregular random bursts are observed in oscillations. With growth of r they become increasingly more frequent until the motion is totally randomized (Fig. 4.5). In the three-dimensional Lorenz model (4.10), the trajectory in the phase space can be computer-estimated.

Figure 4.6 is an example of such a trajectory computed at r=2, σ=10, b=8/3. As seen, the trajectory is attracted to the limited range in the phase space. The motion of the system is wandering, i.e. the trajectory turns irregularly either to the right or to the left. Slight changes in the initial conditions make the new solution deviate from the previous solution with a number of turns. Such behavior is called a strange attractor. The range of

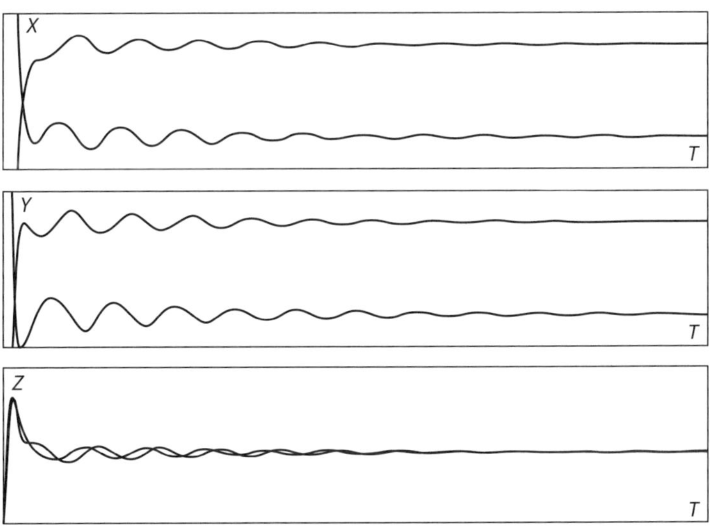

Figure 4.4 Dynamics of variables in the Lorenz model at r corresponding to the presence of two stationary states.

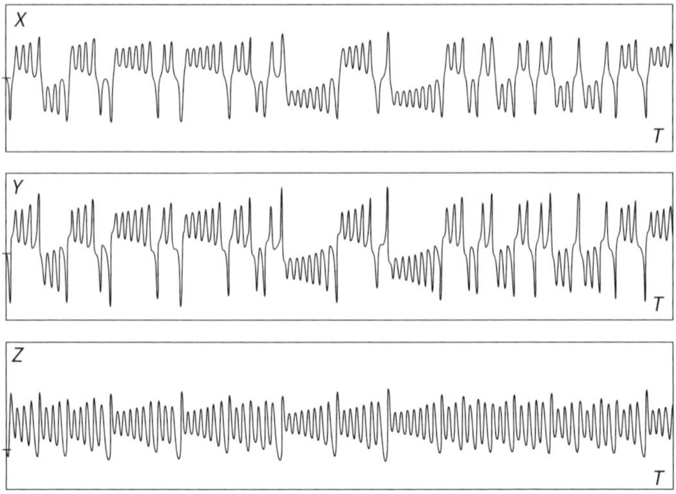

Figure 4.5 Dynamics of variables in the Lorenz model at parameter *r* corresponding to a quasi-stochastic solution.

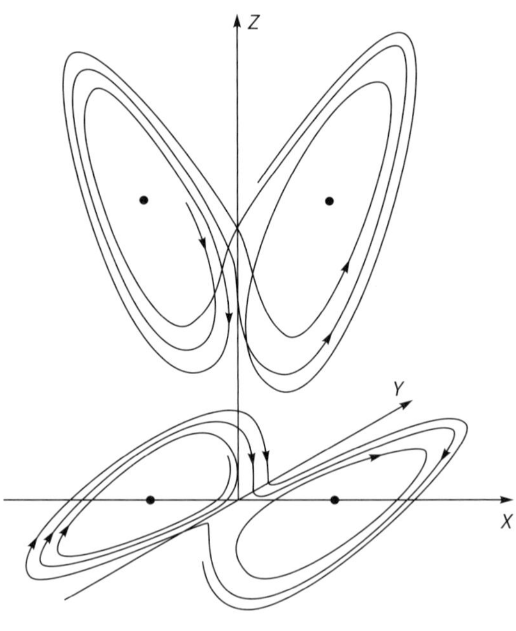

Figure 4.6 Phase trajectories in the Lorenz model. Top, trajectories projected onto the X-Z plane. Bottom, trajectories projected onto the X-Y plane. Points correspond to stationary solutions (according to Lorenz, 1964).

the attractor in the phase space is limited, but it can have a complex structure. The attractor *per se* is formed due to the motion along one trajectory that should pass across every point of the attractor space. But during this process, initially arbitrary close points on the attractor move away from each other at the final distance in a sufficiently long period of time after the beginning of the motion. Apparently, the Lorenz equations may be regarded as a basic model to analyze chaotization of the determined system at a change in the control parameters. The probability of chaos formation in biological systems can be illustrated by the known example of random heart beats at a definite frequency of stimulation pulses.

Model of Population Dynamics. The dynamics of populations in a closed environment may also have random properties. If the population number is not large and at the given moment depends on the population number in previous time periods, the population dynamics can be described discretely using a logistic equation. In the simplest case, after *n* successive generations the population number changes according to the law

$$x_{n+1} = f(x_n) = rx_n(1 - x_n). \quad (4.11)$$

This is a difference equation describing changes in the population number at discrete time moments 0,1, ..., t, t+1, t+2, ... as corresponding numbers in sequence x_0, x_1, ..., x_n, x_{n+1}, x_{n+2}, ..., where every member x_n is dependent on the previous x_{n-1}. Expression (1-x_n) in the right-hand side of (4.11) self-restricts the growth of the population number because of the limited living space that is proportional to (1-x_n); parameter *r* is dependent on the living conditions and fertility.

Function $x_{n+1} = f(x_n)$ obtained from iteration of x_1, x_2, ..., x_n, ... displays complex behavior depending on parameter *r*.

Generally speaking, for fx_n one can observe different regimes: monotonous and oscillatory approach to equilibrium, removal from equilibrium, stable oscillations, and quasi-stochastic behavior (chaos). In equation (4.11), parameter *r* determines the slope of the plot for the function in the right-hand side. At $r < 3$, the population number approaches the stable equilibrium state.

When *r* grows and the plot of the *f*(*xn*) function becomes sharper, bifurcation takes place: stable equilibrium turns into stable cycles (Fig. 4.7). The subsequent growth of parameter *r* in the cycle evokes an increase in the number of periodic points and the periods of new cycles lengthen doubling at every new bifurcation point (Fig. 4.7). With the increasing *r* value, bifurcations occur more frequently, and the population numbers

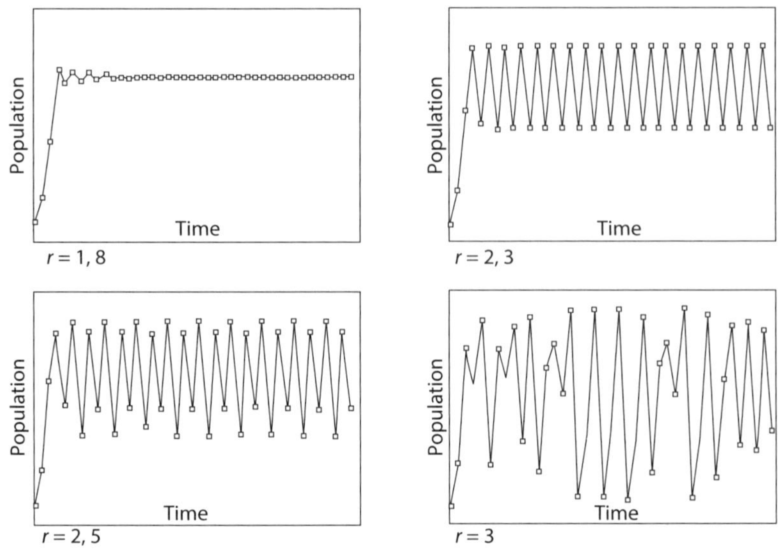

Figure 4.7 Regimes of behavior of variable x in equation (4.11).

recur in every 2, 4, 8, … generations. When r exceeds the critical value $r>r_c=3.56999... \cong 3.570$, the solution is randomized and oscillations become totally random. The value of $r_c=3.570$ is the constant characterizing the threshold of chaotization of the system. At the same time, even at the regime of chaos there exist "windows of regularity", when with the growth of $r > r_c$, stable cycles with periods 3 and 7 are unpredictably resumed. Then bifurcation doubling of periods takes place once more and chaos is revived, that is chaotization takes place via bifurcation with doubling of cycles in the system. It is possible to determine ranges of parameter r wherein every established period remains stable prior to the following increase in r. It has been found that such consecutive values of r when the number of stable periodical points is doubled and becomes equal to 2^n, follow the law

$$r_n = r_c - \text{const}\ \delta^{-n} \text{ at } n >> 1,$$

where $\delta = 4.669... \cong 4.670$ is the Feigenbaum constant. It equals the ratio of consecutive ranges in the r values wherein every cycle retains its stability

$$\delta_n = \frac{r_n - r_{n-1}}{r_{n+1} - r_n}. \tag{4.12}$$

At n $\to \infty$, $\delta_n \to \delta = 4.670$. The value of r_n corresponds to the n-th bifurcation where period 2^n loses its stability and period 2^{n+1} acquires it. As seen from this ratio, the lengths of the ranges diminish in the row of points $r_{n-1} \to r_n \to r_{n+1}$, i.e. doubling occurs more and more frequently with the growth of $r \to r_c$.

Constant δ appeared to be universal being inherent to the behavior of many other natural systems in which doubling of the cycling period takes place before chaos establishes.

General Properties of Deterministic Chaos. In addition to the above described systems, there are a variety of other systems displaying chaotic behavior. The dynamical chaos is understood as irregular random motion in nonlinear systems, for which the general system evolution in time nevertheless obeys the dynamic laws.

The chaotic behavior observed in time originates not as a result of external noise sources, number of degrees of freedom (in model (4.10) they are only three), or quantum-mechanical uncertainty. It is determined by the capability of nonlinear systems with time to "diverge from each other" exponentially rapidly trajectories initially close to each other that at the start were within the limited volume of the phase space. Inasmuch as in reality the starting conditions may be set only with the limited accuracy, the initial error grows exponentially in time so that the real trajectory becomes unpredictable notwithstanding that definite dynamic laws (kinetic equations) control the system behavior. Hence we have the term "deterministic chaos".

Moreover, it has been shown that even in classical mechanics, stable regular motion is rather an exception. At specific parameter values, mechanical motion in the phase space is easily randomized, each time the definite shape of the trajectory being dependent on the starting conditions. A mechanical system of this type is Sinai billiard where the inside table walls are convex so that at low initial deviations of the incident angles of seemingly the same starting trajectories, the consecutive bouncing of the ball from the walls leads to exponential "scattering" of the trajectories and, as a result, to their unpredictability at large time periods.

It should be noted that the setting of the initial state is always uncertain to some degree. But deterministic chaos is found in nonlinear systems when their behavior is sensitive to slight alterations in their initial states. Consequently, the chaotic system obeying deterministic laws behaves chaotically and becomes unpredictable as to the specific type of the trajectories, but it is still quite definite from the point of view of its strange attractor properties as the limited area filled with unstable trajectories.

In Chapter 2 we have described an autooscillatory system of glycolysis with the experimentally detected intermittent changes in intermediate concentrations (NADH, PPK, ADP, F6P, and FDP) in yeast cells.

The model of autooscillations in glycolysis (Chapter 2) has two equations (2.23) the analysis of which demonstrates the existence of a stable limit cycle as a regular dynamic attractor of this point system. Yet it may be expected that in such a complex system as glycolysis, even more complex dynamic behavior, including chaotic behavior, may be observed under certain conditions consistent with the addition to model (2.23) of a third equation. As a matter of fact, direct experiments on yeast cells at continuous regular addition of glucose allowed to register aperiodic stochastic alterations in NADH concentrations detected by the changes in the fluorescence of this compound.

It was revealed that more frequent glucose injections give rise to chaotic oscillations. The scenario of chaos formation in a real system is quite complex. In this case, chaos establishes with the doubling of the number of periodic points, the doubling of the cycling period in bifurcation points, and the emergence of "windows of regularity" at chaotic regimes.

Figure 4.8 shows experimentally observed aperiodic chaotic changes in the fluorescence of NADH in cells of yeast extracts with sinusoidal glucose inflow.

Hypothetically, the presence of chaotic regimes in itself may both be associated with distorted metabolism and characterize the transient state of metabolic processes that is formed in response to external effects.

Thorough studies reveal chaotic behavior in a number of dissimilar systems in which the chaotic state emerges in real conditions and plays a vital role in the dynamics. Such are examples of irregular bursts in the population number, unpredictable peaks of incidence rates during mass

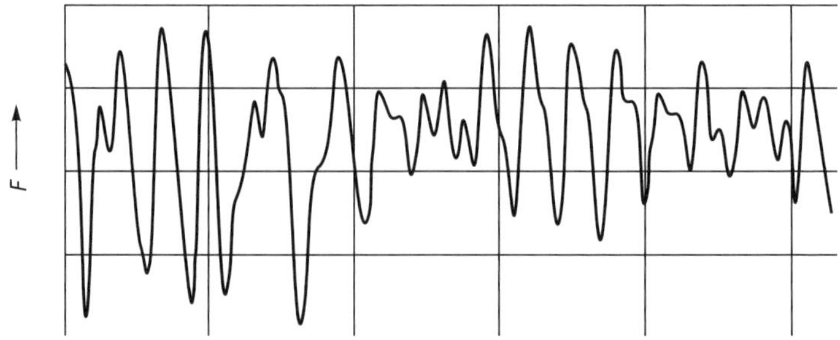

Figure 4.8 Measured fluorescence of NADH in yeast extracts at sinusoidal inflow of glucose (M. Marcus et al., Biophysical Chemistry, 1985).

epidemics that in no way correlate with external conditions and methods of treatment; cardiac dysrhythmia and cardiac fibrillation; chaotic movements of eye muscles at schizophrenia; chaotic small-scale intramolecular movements of the white of the eye. From this point of view, of special attention is also the problem of morphogenesis in nature and in particular in biological systems. Thus, in the process of morphogenesis the genetic program contained in the DNA predetermines only the sequence of biosynthesis of macromolecules. Then they interact in the cell obeying the dynamic regularities and giving rise to the emergence of dissipative structures including fractal structures.

There are well-known examples of the branching structure of different organs and tissues (air vesicles in lungs, dissected leaves) consisting of randomly compiled small details, which when combined nevertheless preserve specific contours inherent to the whole body in general.

It appeared that for computer imitation of a complex fractal structure there exist quite simple rules following which it is possible to reproduce its formation according to the laws of chaos. The chaotic behavior is in fact a reflection of profound regularities of the dynamic organization of complex systems. The models of deterministic chaos proposed herein may embody the simplest rules of chaotization and as such may be regarded as original basic models of chaos. It is evident that here we are only at the beginning of learning the role of chaos in nature and in self-organization of biological systems.

At present the main general result is that, given certain parameters, the behavior of deterministic systems, including biological systems that have been always thought as predictable, in effect may have chaotic properties.

5

Mathematical Models in Ecology

The solution of current problems in ecology is not only of cardinal scientific significance but of great importance to mankind. Of special meaning is investigation of the dynamics of ecological processes aimed at elaboration of methods for predicting the development of ecological systems and their optimal regulation for the sake of humans. One of the important directions in this field is mathematical modeling of processes of energy and mass exchange in biogeocenosis which involve biomasses of links in ecological systems.

Basic models of the dynamics of population numbers in ecosystems will be analyzed and it will be demonstrated how they can be used to characterize the stability of ecological communities and predict their behavior in time.

Models of a Single Population. Let us consider the behavior of a single-species population without explicit considering its interactions with other species located at the same or another trophic level. The simplest equations describing the unlimited population increase in conditions of redundancy of food and space (for example, penicillin fungi in a cultivator) have been already discussed in Chapter 1 (1.1–1.3).

It should be reminded that equation (1.3) for the Verhulst logistic curve is obtained with the account for self-limitation of growth ("self-poisoning") of the population in conditions of crowding and competition within the population. It reads as follows

$$\frac{\partial x}{\partial t} = \varepsilon x - \delta x^2, \frac{\partial x}{\partial t} = \varepsilon x - \delta x^2, \tag{5.1}$$

where δx^2 is proportional to the number of encounters of the species within the population and takes into account this effect of growth self-limitation.

It is seen that the equation of stationary states

$$\varepsilon \bar{x} - \delta \bar{x}^2 = 0$$

has two roots

$$\bar{x}_1 = 0,$$

$$\bar{x}_2 = \frac{\varepsilon}{\delta}.$$

In accord with the stability criterion (1.17), the derivative sign in the right-hand side of (5.1)

$$f'(x) = \varepsilon - 2\delta x \tag{5.2}$$

defines its stability in the stationary point. Substituting $\bar{x}_1 = 0$ for the first stationary regime, we obtain $f'_{\bar{x}_1=0}(x) = \varepsilon > 0$, i.e. the state where $\bar{x}_1 = 0$ is unstable.

On the contrary, the other stationary state $\bar{x}_2 = \frac{\varepsilon}{\delta}$ is stable because in all cases

$$f'_{\bar{x}_2}(x) = -\varepsilon < 0.$$

The stable state $\bar{x}_2 = \varepsilon / \delta = \bar{x}_{max}$ corresponds to the maximal stationary population number or the "environmental capacity" admissible in the specified conditions. As seen, this value depends on the relation of constants ε and δ of the "inflow" and "diminution" processes in the population number. It is expedient to write equation (5.1) like this

$$\dot{x} = r x(1 - x / k) \tag{5.3}$$

where $r = \varepsilon$, $k = \dfrac{\varepsilon}{\delta}$.

Equation (5.3) is a simple mathematical model and by studying its properties it becomes already possible to understand the effect of certain factors on the population fate.

As seen from (5.1) and (5.3), an increase in the population number x is dependent on the reproduction rate as εx. Meanwhile this linear expression is valid only for asexual reproduction (parthenogenesis). For a heterosexual population, under conditions of unlimited resources the reproduction rate is determined by the number of encounters of males and females and grows quadratically with the rise in the species number

$$\dot{x} = r x^2 . \tag{5.4}$$

However given the population density is very high, the reproduction rate is limited by the number of females in the population rather than by the number of encounters of males and females. The both effects (the quadratic dependence of the reproduction rate and its specific limitation at high population members) are registered as

$$\dot{x} = a \frac{\beta x^2}{\beta + \tau x} . \tag{5.5}$$

Formula (5.5) does not limit the population number and does not predict its downfall at low population densities. But in actual ecological conditions the population is observed to perish if its number drops below some critical value. Evidently this is caused by the fact that at low population densities, the time required for encounter and fertilization is already longer than the lifetime of an individual species capable of fertilization. Take this factor into account by introducing in (5.5) a member describing the death rate proportional to the population number

$$\dot{x} = a \frac{\beta x^2}{\beta + \tau x} - \gamma x . \tag{5.6}$$

This equation has two stationary solutions

$$\bar{x}_1 = 0, \quad \bar{x}_2 = \frac{\gamma \beta}{a \beta - \gamma \tau} = l. \tag{5.7}$$

The analysis shows that point $\overline{x}_1 = 0$ is stable, and point $\overline{x}_2 = l$ is unstable. This should be interpreted as follows. If the initial population number is lower than $l(x < l)$, then the population degenerates and its number vanishes ($x \rightarrow 0$). But in accordance with (5.6) at $x > l$ the population will grow infinitely. To predict the existence of population of definite species, it is necessary to know the lower critical value for its number below which the population would inevitably perish. For example, it has appeared that the population of blue whales is doomed to perish for at present its density is below the critical value. And this is true even notwithstanding that individual species can be found in the World Ocean and hunting for blue whales is banned. It is also clear that the actual population number should be restricted from above as at high densities the reproduction rate drops (member δx^2 in 5.1) because of the intraspecies competition. The expression that takes into consideration the both factors – the lower critical boundary of the population number and self-limitation at high densities – reads like this

$$\dot{x} = a\frac{\beta x^2}{\beta + \tau x} - \gamma x - \sigma x^2. \tag{5.8}$$

Figure 5.1 shows dependencies of the population number on time (*a*) and rate $\dot{x}(t)$ versus $x(t)$ (*b*).

System (5.8) has already three stationary points (as a result of the emergence of cubic members $(-\tau\delta x^3)$): $\overline{x}_1 = 0$ and $\overline{x}_3 = K$ are stable points,

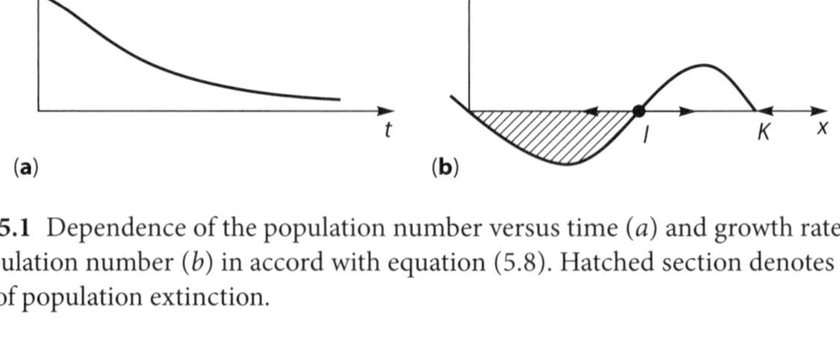

Figure 5.1 Dependence of the population number versus time (*a*) and growth rate versus the population number (*b*) in accord with equation (5.8). Hatched section denotes the region of population extinction.

and $\overline{x} = l$ is an unstable point positioned on the boundary of ranges of the two stable regimes (K and l are functions of equation 5.8). Subject to its initial number, the population either perishes ($x \rightarrow 0$ at $x_{ini} < l$) or reaches its maximally admissible value ($x \rightarrow K$ at $x_{ini} > l$).

Logically, in trying to predict the destiny of a specific population it is necessary not only to know *a priori* its initial population number, but also to determine through observations and experiments the K and l values. In some cases, it works well to forecast that the population is approaching the dangerous "no-return" boundary. Below this boundary it is already impossible to restore population number and that is of special importance for planning fisheries. If in response to a single-time act of reduction in the number of a well developed population it is restored rapidly (Fig. 5.1, curve 1) without any initial delay, the critical point ($\overline{x} = l$) is yet distant. And to the contrary, the emergence of a lag phase in the process of restoration (Fig. 5.1, curve 3) evidences to the dangerous proximity of the boundary of irreversible reduction in the population number and hence to the need of diminishing the fishery scales.

Effects of Delay. In actual ecosystems, the processes of reproduction and death occur at different moments of time. This means that in equations, the rate of changes in the population number should depend on the number at the preceding moment of time rather than at the same one. In other words, the response of the system to the changes in its variables takes place not immediately , but after some time interval T (i.e. is delayed). In ecosystems this is associated first of all with the period of child-bearing and organism development. If the time of an adult species development is T, the equation generally describing the dynamics of the number of adult species

$$\dot{x} = f(x),$$

should be substituted by the following

$$\dot{x} = f(x_{t-T}), \tag{5.9}$$

where x_{t-T} is the number of eugamic species at moment $t - T$.

Another effect of delay in the regulation of the population number is associated with the seasonal character of reproduction of various species. In the case adult species reproducing in the given year do not survive till the reproduction period in the next year the "annual" delay equation will be as follows

$$\dot{x}_{n+1} = f(x_n), \tag{5.10}$$

where n is the population number in the n-th year.

Self-limitation of the population number taken into account in equation (5.1) as the quadratic member ($-\delta x^2$) is a result of both poisoning the environment with metabolic products and cannibalism. These factors depend first of all on the number of adult species and have the maximum effect on the early-age stages. This means also that at the specific moment, restriction of the general number x depends on the impact on the species population at the previous moment. With account of these phenomena, logistic equation of growth (5.1) can be rewritten as

$$\dot{x} = \varepsilon x - \delta x x_{t-T} = x(\varepsilon - \delta x_{t-T}) \tag{5.11}$$

or instead of equation of growth (5.3) we get the Hatchinson equation well known in mathematical ecology

$$\dot{x} = rx\left(1 - \frac{X_{t-T}}{K}\right) \tag{5.12}$$

where T is the age of maturity.

Solutions of delay equations reveal a variety of dynamic regimes in ecosystems and, in particular, show conditions for emergence of oscillations in the population number. Here the general reason for the appearance of oscillations is the same as in the system regulated with the feedback loop in which the "limit signal" is essentially delayed. Just this happens in the population where, at the given moment t, self-limitation of its number x occurs with time delay T at the cost of life activity of adult organisms at the preceding moment $t - T$. This is particularly displayed in systems where the delay in the feedback loop exceeds the characteristic time of the system itself. In ecosystems (5.11) and (5.12) at $T >> 1/\varepsilon$ ($1/\varepsilon$ is the characteristic time of the system), divergent/rising oscillations may appear, though the equation without delay does not give any oscillations in the population number (see Fig. 5.1). Depending on parameters, oscillations can be regular periodical or, quite the reverse, chaotic when their frequency and amplitude are not constant. Such properties are characteristic of some model and laboratory ecosystems.

Figure 5.2 shows the experimental curve of oscillations in the population number of flies and larvae in a population box. The analysis of a corresponding model (Fig. 5.3) demonstrated that instability and divergent/rising oscillations appear only when the time of development from egg

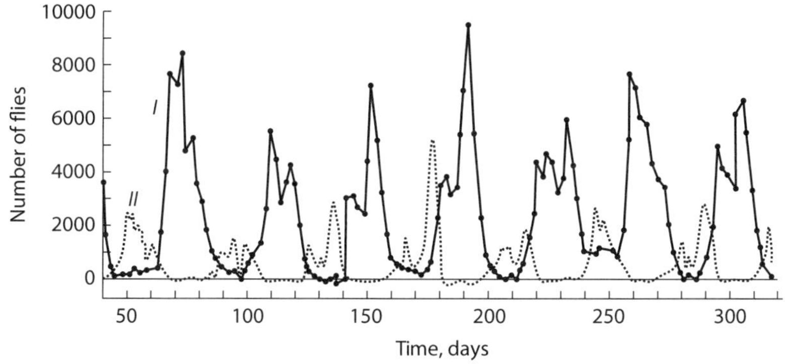

Figure 5.2 Population number of *Lucilia C.* flies in a population box (Nicolson, 1954). I, number of adult species; II, number of eggs laid a day.

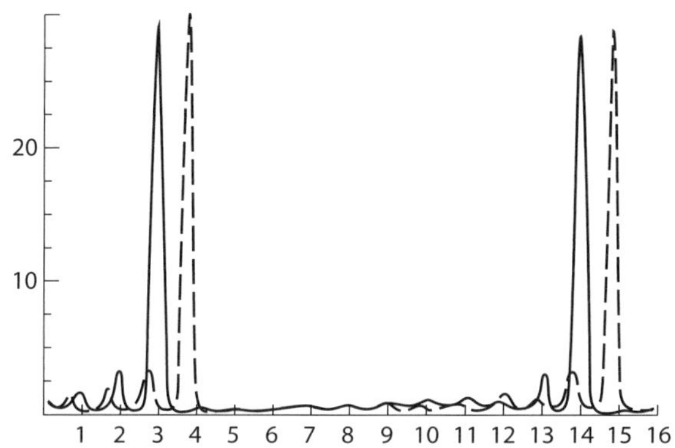

Figure 5.3 Dynamics of population number in the delay model of larva-imago: quasi-stochastic behavior.

to adult species (delay) is larger than the time of natural death $(T > 1/\varepsilon)$. For other models of chaotic oscillations it was found that the amplitude of a regular burst in the population number is directly proportional to the interval between the bursts (the case of regular chaos). Probably the existence of superbursts of insects in nature suggests that the dynamic regime of such populations is "regular chaos".

Models of Species-to-Species Interaction. The vast variety of modes of species-to-species interactions might have made improbable generalized models which would reflect the regularities in population growth in nature. Nonetheless depending on whether the population number of one species type increases or remains unchanged in the presence of another type, it is possible to classify the interactions without going into details of their mechanisms. Accordingly, the models of species-to-species interactions can be of three types:

1. species competition leading to a decrease in the population numbers of both types;
2. predator-prey interactions when an increased number of predators (or parasites) occurs because of a decrease in the number of prey (hosts);
3. symbiosis leading to an increase in the number of both types.

Suppose as it was done on the example of the Volterra predator-prey model, that alteration in the population number (biomass) of one of the two types is proportional to the probability of encounter of their species, i.e. to the product of their population numbers. So the equations describing the species-to-species interactions, the population number of each of them being x_1 and x_2, respectively, are rewritten like this

$$\dot{x}_1 = c_1 x_1 - a_{11} x_1^2 + a_{12} x_1 x_2 \,,$$
$$\dot{x}_2 = c_2 x_2 + a_{21} x_1 x_2 - a_{22} x_2^2 \,. \tag{5.13}$$

Similar to (5.1), in these equations linear members c_1x_1 and c_2x_2 describe unlimited reproduction of species, and members $(-a_{11}x_1^2)$ and $(-a_{22}x_2^2)$ comply with self-limited growth of their population number. The appearance of the $a_{12}x_1x_2$ and $a_{21}x_1x_2$ members is caused by species-to-species interaction the nature of which is taken into account by the sign of coefficients. Given the species compete with each other the coefficients would be negative $a_{12} < 0$, $a_{21} < 0$. Upon symbiosis the both coefficients are positive ($a_{12} > 0$ and $a_{21} > 0$). In the case of the predator-prey interactions, coefficients a_{12} and a_{21} have opposite signs.

The analysis of system (5.13) for its stability is performed taking into consideration the values and signs of coefficients in the equations. We will not go beyond examination of some basic results. In conditions of competition, equations (5.13) will look like this

$$\dot{x}_1 = x_1 (c_1 - a_{11} x_1 - a_{12} x_2) \,,$$
$$\dot{x}_2 = x_2 (c_2 - a_{21} x_1 - a_{22} x_2) \,. \tag{5.14}$$

By equating the right-hand sides to zero we get that the above system has four stationary points, the character of stability of them being able to alter depending on the values of coefficients. The analysis demonstrates that only one point corresponds to the stable existence of the two types of species provided the ratio of coefficients is valid

$$a_{12}a_{21} < a_{11}a_{22}. \tag{5.15}$$

In all other cases, the points are either unstable, or steady regimes are observed only at the zero population number of one of the species, that is when a particular species survives.

Inequality (5.15) means that the competing species can coexist in a stable system when the product of coefficients of interpopulation interaction $a_{12}a_{21}$ is lower than the product of coefficients of intrapopulation "self-limited" interaction a_{11},a_{22}. This can be explained if we suggest that suppression of the population number of one of the competitors takes place mostly because of its own self-limited growth rather than because of the growth of another competitor. In other words, the extremely enlarged population itself limits its growth, thus enabling the neighboring population of competitors to survive. Naturally, this can be observed in relatively favorable conditions. Whereas in harsh conditions when resources are limited and "all energy is spent on fighting the enemy", the population growth is limited and coefficients a_{11} and a_{22} are low as compared to a_{12} and a_{21}. Hence inequality (5.15) is violated and the stable regime becomes plausible only upon survival of the toughest competitor. The predator-prey interactions have been analyzed above by the example of the Volterra model (see Chapter 2), where it is shown that depending on the ratio of parameters, the singular point of system (5.16)

$$\begin{aligned} \dot{x}_1 &= x_1(c_1 - a_{11}x_1 - a_{12}x_2), \\ \dot{x}_2 &= x_2(c_2 - a_{21}x_1 - a_{22}x_2). \end{aligned} \tag{5.16}$$

may be either a steady node (at large a_{11} and a_{22}) or a steady focus. In a particular case when coefficients a_{11}=0 and a_{22}=0, the system has a singular point such as the center (see Fig. 2.8). This model becomes structurally unstable and does not generate undamped oscillations because at any random fluctuation in the population number the system switches from one ellipsoidal trajectory to another thus changing the oscillation amplitude. But both in natural and experimental conditions, stable oscillations of the population number can be observed, and therefore the system should have regulatory mechanisms for maintaining the oscillatory state which are not taken into consideration in the initial model (5.16).

Let us note that mathematically stable oscillations in the system are plotted on the phase plane as trajectories of the limit cycle described in Chapter 2 (see Fig. 2.12) by the example of oscillations in a glycolytic cycle (see Fig. 2.13). The problem is how to select corresponding mathematical functions in order to reflect in models the role of biological factors of population interaction that are not taken into account in the simplest Volterra model. Then it is necessary to determine the character of stability, including the presence of autooscillations and the limit cycle, depending on the ratio of parameters in the equations.

As a result of the model analysis it is possible to determine peculiarities of the dynamic behavior of the system that can be of general biological meaning. Crucial biological factors not taken into account in the Volterra model (5.16) still regulate the rate of prey consumption in predators and satiety of predators with food, their competition for the prey, reproduction and death rate of prey and predators as well as other limiting mechanisms. Let us name some of them. Thus, in model (5.13) the expression for the rate of prey extinction by predators ($a_{12}x_1x_2$ members) has no indication for predator satiety which is observed only at low prey densities. At high prey density, we have satiety that can be described by the function of extinction or trophic function φ(x) that reads like this

$$\varphi\,(x_1) = \frac{\varphi_{\max} x_1}{1 + a x_1}, \tag{5.17}$$

where φ_{max} is the predator maximal ration when food is profuse, α is the prey population density (α is constant such that 1/α is the prey population density at which ration φ(x) makes half of the maximal, i.e. $\varphi\,(x_1)=0.5\varphi_{max}$). Then the rate of prey extinction will be expressed as $\varphi(x)x_2$. It is suggested to describe another effect of competition of predators for prey using a mathematical function that also reflects the state of satiety. In other words, instead of $\varphi(x_1)x_2$ the expression for the rate of prey extinction now has the following member

$$\varphi\,(x_1)\cdot\frac{b x_2}{1+\beta x_2} = \frac{\varphi_{\max}\cdot x_1}{1 - a x_1}\cdot\frac{b x^2}{1+\beta x_2} \tag{5.18}$$

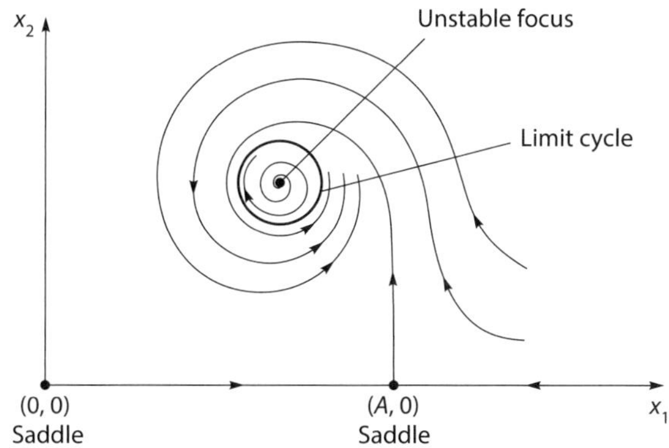

Figure 5.4 Phase portrait of a system describing interaction of two species (Kolmogorov, 1972). Closed trajectory (limit cycle) is shown.

that, as seen, differs greatly from $a_{12}x_1x_2$ in the initial Volterra model. There are also many other mathematical formulas for functions of reproduction, death and extinction, which are used for describing different biological situations.

It is obvious that depending on the type of functions, the right-hand sides of the equations may show various dynamic behavior of the system. Thereby general questions have arisen: What types of functions can be used to describe the dynamics of the population number of interacting predator–prey populations, and what specificity will be observed in phase portraits of the models depending on their properties? We will not go into mathematical details but will give the result of studies (Kolmogorov, 1972). It was found that at different proportions of parameters, the system could have two or three special points. One of them lies at the origin of the coordinates $\bar{x}_1 = 0, \bar{x}_2 = 0$ and is a saddle in all cases. The other two could be either a saddle or a stable or unstable focus and node. If the stationary point is an unstable focus it can be surrounded by limit cycles, i.e. stable regular oscillatory solutions (Fig. 5.4). One of the designed models describes the dynamics of the predator-prey population number with account for the effects of satiety of predators and intraspecies competition of prey and predators. After substituting the variable the system is as follows

$$\dot{x}_1 = x_1 - \frac{x_1 x_2}{1 + a x_1} - \varepsilon x_1^2,$$

$$\dot{x}_2 = -\gamma x_2 + \frac{x_1 x_2}{1 + a x_1} - \mu x_2^2. \qquad (5.19)$$

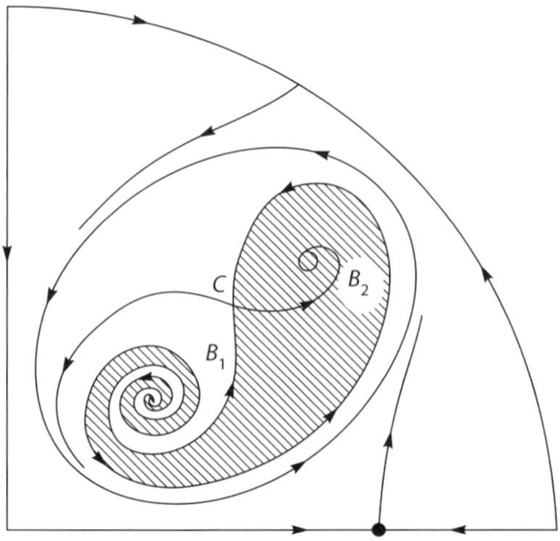

Figure 5.5 Example of a phase portrait of system 5.19 (see the text).

This model has a set of phase portraits depending on the combination of parameter values. Figure 5.5 shows an example of a phase portrait when the system has three stationary points: B_1 is an unstable focus; C is a saddle; and B_2 is a stable focus. These points are encircled by the trajectory of a stable limit cycle.

The system in unstable region B_1 may with certain probability move to stable region B_2 or to the stable limit cycle. In real systems, the change of the regimes accompanied by crossing over the bifurcation boundaries and transition from one stable region to the other takes place at rather tough external effects resulting in alteration of the species population number.

Stochastic Population Models. The above population models are deterministic. However in real life a system may be affected by random effects, connected with fluctuations in the species number or parameter values of the system. Besides, as a matter of fact, the very processes of reproduction and death are of probabilistic nature. At a large number of species, the deterministic description is compatible with the stochastic, i.e. the data on the species number obtained with solutions of differential equations coincide with the corresponding mathematical expectations. But the accounting of the stochastic nature of ecological processes becomes especially important at small population numbers. In such a case, the mean square deviation in the number of a particular species from the mathematically expected one may be quite noteworthy. This leads to essential oscillations in the population plot displayed upon examining a particular population which characterizes the fluctuating changeability of the given process and its deviation from the theoretical curves (phase trajectories) specified by the deterministic model. For example, if at some point of the phase trajectory variable (x_2) is not very large, then chaotic fluctuations may cause the representation point to move from the phase trajectory to one of the axes (axis x_1), that is the population number of the species (x_2) becomes zero and this species (x_2) perishes. So, in the long run the stochastic model can predict the extinction of one of the species. It should be underlined once again that such effects are revealed at small population numbers.

Effects of Spatial Organization. In Chapter 4 we analyzed models with distributed parameters in which variables are changed not only in time but in space as well. Let us accept that migration of both predators and prey in space has the character of random wandering analogous to diffusion. Then the behavior of a simple Volterra system can be described using equations similar to (4.6)

$$\frac{\partial x_1}{\partial t} = c_1 x_1 - a_{12} x_1 x_2 + D_1 \frac{\partial^2 x_1}{\partial r^2} \quad ,$$

$$\frac{\partial x_2}{\partial t} = a_{12} x_1 x_2 - c_2 x_2 + D_2 \frac{\partial^2 x_2}{\partial r^2} \, . \tag{5.20}$$

Here D_1 and D_2 are diffusion (migration) coefficients for species in their population range, and self-limitation effects (members $a_{11}x_1^2$ and $a_{22}x_2^2$) are absent. In this system with a limited population range, the qualitative pattern of regular oscillations is maintained. But if the population range is not limited, that is the system is not closed in space, it may generate solutions as propagating waves. The analysis of model (5.20) was performed on the assumption that $D_1 = 0$, i.e. prey migration is absent, which in a real situation evidences of an essentially lower mobility of prey as compared to that of predators. The obtained solutions have the shape of prey and predator waves propagating in space.

Figure 5.6 demonstrates the distribution of population density of prey (φ_1) and predators (φ_2) along the spatial coordinate r at a fixed time moment. It is the so-called "pursuit and evasion wave". With time this wave is propagating along coordinate r. Naturally, each time at the starting point these waves are regenerated on account of the point model equations consistent with the active character of distributed system (5.20).

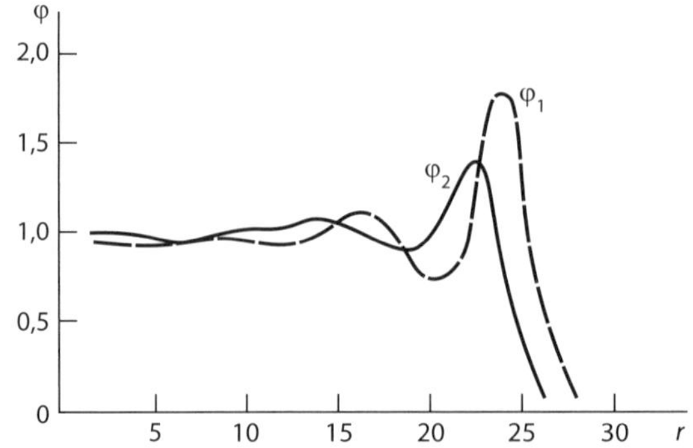

Figure 5.6 Distribution of population densities of predators (φ_2) and prey (φ_1) in space.

Another type of the original spatial behavior of ecological systems is stationary nonhomogeneous distributions of variables in space – dissipative structures that can be compared to "natural life stains". Of special importance is the question: How are autooscillatory regimes of point models and dissipative structures interconnected in appropriate distributed systems?

One of the simple point Volterra models with a limit cycle is represented as a suitable distributed system like this

$$\frac{\partial x_1}{\partial t} = a x_1 \left(\frac{k - x_1}{x_1} \right) - b x_1 x_2 + D_1 \frac{\partial^2 x_1}{\partial r^2}$$
$$\frac{\partial x_2}{\partial t} = -c x_2 + \alpha x \tag{5.21}$$

In this system as compared to the Volterra model, member $a x_1^2 \left(\frac{k - x_1}{x_1} \right)$ describing the dynamics of prey in the absence of predators has cubic nonlinearity (cf. (5.3)). This is explained by sexual reproduction of prey when at low densities the growth rate is proportional to the number of encounters of species, i.e. to the squared population density (ax_1^2).

System (5.21) was computer analyzed in a numerical experiment under condition of low prey mobility ($D_2/D_1 = 1000$) and such a proportion of parameters, when the point system generates stable autooscillations (α/c=1 and c/αK=0.4). The system was found to have two different regimes. At some initial conditions the autooscillatory regime is formed when

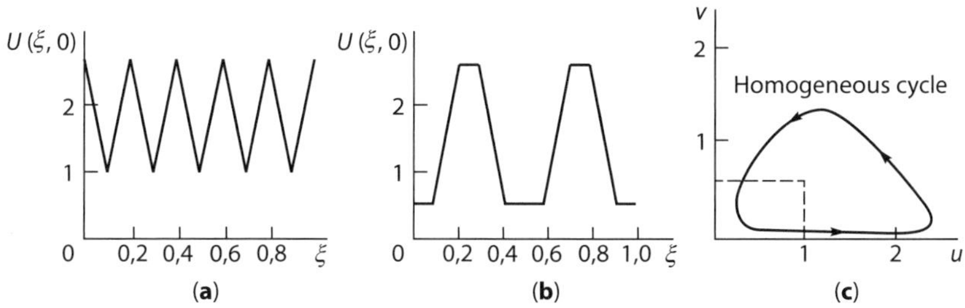

Figure 5.7 Initial distributions of prey population density (*a* and *b*) leading to synchronous homogeneous oscillations over the entire ring range (*c*) (Bazykin and Markman, 1980).

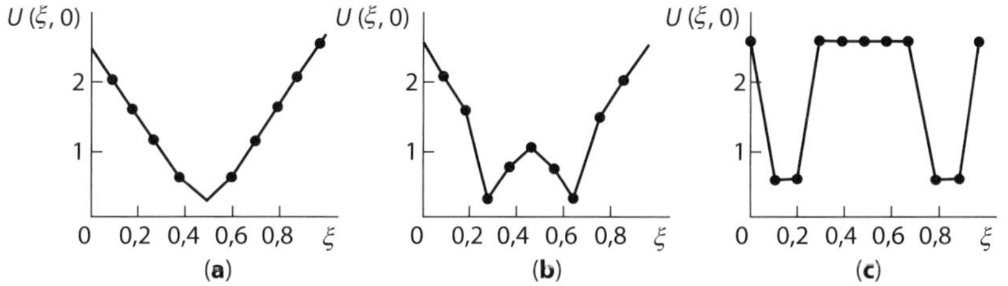

Figure 5.8 Initial distributions of prey population density (*a*-*c*) leading to the formation of a stationary dissipative structure.

components are distributed in space homogeneously. This is consistent with a stable limit cycle of point system (Fig. 5.7) when synchronous oscillations in the population number are generated over the entire population range.

There are also other initial conditions when with time the system forms a stable dissipative structure, i.e. a stable regular distribution of concentrations in space (fig. 5.8). In this case, the distribution of predators rapidly migrating in the population area is close to homogeneous (they get "smeared" in the system). On the contrary, the concentration of slowly migrating prey with a settled way of life differs at various spatial points.

6

Thermodynamics of Irreversible Processes in Biological Systems Near Equilibrium

In general classical thermodynamics considers equilibrium states of a system wherein system parameters and characteristics do not change with time. The actual transitions between different states are not among the subjects of classical thermodynamics, and its methods permit evaluating only typical energy effects of chemical transformations by comparing parameters of the initial and final states of the system. However in open systems, reactions and appropriate energy transformations occur regularly, therefore it is required to know the rates of energy transformation processes at every time moment. This means that in calculating energy effects it is necessary to take into account the time factor as well, thereby to somehow combine thermodynamic and kinetic approaches in describing the properties of an open system. We will analyze thermodynamic criteria of stability of stationary states and criteria of attaining them close to and far from equilibrium and will become acquainted with thermodynamic characteristics of autooscillatory and trigger regimes.

The First and Second Laws of Thermodynamics. Let us remind the basic laws of classical thermodynamics and results of their application in biology. According to the first law, the amount of heat δQ absorbed by the system from the external environment is spent for increasing its internal energy dU and performing the overall work δA that includes both work against the external pressure P to change the volume dV of the system and the effective work $\delta A'_{\text{мax}}$ accompanying chemical transformations:

$$\delta Q = dU + \delta A$$

where the work is expressed as $\delta A = pdV + \delta A'_{max}$
or

$$\delta Q = dU + pdV + A'_{max}. \tag{6.1}$$

The experimental checking of the first law was performed in special calorimeters used to measure the heat generated by an organism in metabolism processes, upon evaporation and also in secreted products. The nutrients received by organisms are decomposed with the release of their free energy used for life activity. Experiments were carried out on the set of reactions of basal metabolism at a relatively short period of time when biomass accumulation does not take place as a result of growth and no noticeable work is performed. It was found that the heat released by the organism totally complies with the energy absorbed via nutrient consumption. The validity of the first law signifies that the organisms themselves are not an independent source of any new form of energy.

The second law of thermodynamics confers the criterion of the direction of spontaneous irreversible processes. Any alteration in the state of the system can be described by the corresponding change in the special function of the state – entropy S which is determined by the total value of normalized heats Q/T absorbed by the system.

For equilibrium processes the small change of entropy dS is equal to elementary normalized heat $\delta Q/T$ absorbed by the system and is larger for non-equilibrium processes

$$dS \geq \delta Q/T. \tag{6.2}$$

In isolated systems $\delta Q = 0$ and hence

$$dS \geq 0. \tag{6.3}$$

This is the evolutionary criterion of direction of irreversible changes in isolated systems which in all cases proceed with an increase in the entropy to its maximal values upon termination of the process and formation of thermodynamic equilibrium. The entropy increase testifies to a decrease in the level of ordering and organization of the system, i.e. its chaotization.

We will be interested in the relation between the thermodynamic parameters and the value of the maximally effective work $\delta A'_{max}$ which characterizes internal irreversible chemical transformations in the system. It is impossible to determine univalently the $\delta A'_{max}$ value using expression (6.1) for the first law upon transition from one state to the other because the δQ value depends on the transition pathway. But by combining expressions (6.1) and (6.2) one can find special characteristic functions, their change being equal to $\delta A'_{max}$ in definite conditions. In biochemical processes, the most important are free energy F

$$F = U - TS$$

and total thermodynamic potential G or Gibbs energy

$$G = U + pV - TS.$$

If the processes occur at constant T and V, then

$$\delta A'_{max} \leq - \mathrm{d}(U - TS) = \mathrm{TdS} - \mathrm{dU} = - (dF)_{T,V},$$

and if T and p are constant we have

$$\delta A'_{max} \leq - \mathrm{d(U+pV - TS) = TdS - dU - PdV} = - (\mathrm{dG})_{T,P}, \quad (6.4)$$

where the inequality sign complies with irreversible processes.

So, the performance of effective work $\delta A'_{max}$ entails a decrease in the free energy and Gibbs energy in irreversible processes. Note that typically in biochemical transformations the change in the system volume can be neglected; $dV \approx 0$ and therefore the dF and dG values are compatible. Classical thermodynamics allows one to calculate energy effects and accordingly determine the direction and possibility of coupling different biological processes.

There are various methods for estimation of values ΔG and ΔF that are described in detail in physical chemistry. So, given the equilibrium constant (K) of a chemical reaction identified, it is possible to estimate the ΔG_0 value, that would comply with the decrease in ΔG upon transition from the initial non-equilibrium state of the mixture, in which initial concentrations of its components are equal to unity, to the final equilibrium state

$$\Delta G_0 = - RT \ln K,$$

where R is the gas constant (1.987 cal/K/mol or 8.314 J/K/mol). For example, in the reaction

$$\text{Glucoso-1-phosphate} \rightleftarrows \text{Glucoso-6-phosphate}$$

equilibrium constant $K = 17$, hence it follows that $\Delta G_0 = 17000$ cal/mol < 0. The negative value $\Delta G < 0$ shows that in standard conditions this reaction is spontaneous and is essentially shifted to the right, which is supported also by the high value K = 17 >> 1.

Similarly, hydrolysis of ATP that proceeds with splitting of the residue of phosphorous acid and its transfer to water

$$\text{Adenosine} - \text{Ph} \sim \text{Ph} + H_2O \rightleftarrows \text{Adenosine Ph} \sim \text{Ph} + H_3PO_4,$$

is characterized by negative value $\Delta G_0 = -7$ kcal/mol. It is the release of a relatively large portion of energy during hydrolysis of ATP that makes the Ph ~ Ph bond in ATP a macro-energy bond as compared to other reactions of group transfer.

By comparing ΔG_0 values of different processes one can determine whether their coupling is possible when one (coupled) process occurs with the increase in ΔG at the expense of the ΔG reduction in the other (coupling) process. So, oxidation of a glucose molecule during respiration is accompanied by a decrease of $\Delta G_0 = -678$ kcal/mol. It is equal to the increase in ΔG_0 during photosynthesis upon the formation of a glucose molecule from water and CO_2. Thereby from the thermodynamic point of view, coupling of photosynthesis and respiration is possible. The same approach is used when the aim is to find the possibility of coupling other simpler processes as well. The formation of acetyl phosphate in the phosphorylation reaction

$$\text{Acetyl CoA} + \text{Phosphate} \rightleftarrows \text{Acetyl phosphate} + \text{CoA}$$

proceeds with an increase in $\Delta G_0 = 3$ kcal/mol > 0 and can be readily realized upon coupling to ATP hydrolysis or the reaction

$$\text{Glucoso-1-phosphate} + H_2O \rightleftarrows \text{Glucose} + \text{Phosphate}$$

where $\Delta G_0 = -4.8$ kcal/mol.

It should be mentioned that in all the above cases, one can bear in mind only the thermodynamic probability of the process rather than its actual occurrence by the given molecular mechanism. Moreover, the direct character of the temporal change of the free energy in the course of the reaction or upon coupling of different processes is not taken into consideration with such an approach.

The driving force of a chemical process is associated with the difference in chemical potentials of initial and final products. If this difference vanishes, the process is terminated and equilibrium is formed in the system when the rates of the direct and reverse reactions are equal. For example, in the simple case of a monomolecular reaction where the stoichiometric coefficients are equal to unity

$$C_1 \rightleftarrows C_2,$$

the driving force A merely equals the difference of chemical potentials μ_{c_1} and μ_{c_2} :

$$A = \mu_{c_1} - \mu_{c_2}.$$

Here $\mu_{ci} = \mu_{o,ci} + RT1nC_i$ is the chemical potential of substance C_i. The A value is also called the chemical affinity of the reaction. It may be represented as

$$A = -\sum v_i \mu_i, \tag{6.5}$$

where v_i values are stoichiometric coefficients in the equation for the reaction

$$v_1 C_1 + v_2 C_2 + \ldots \rightleftarrows v'_1 C'_1 + v'_2 C'_2 .$$

They are included in equation (6.5) with the negative sign for substances from the left-hand side and with the positive sign for substances from the right-hand side of the reaction equation. At equilibrium when all $C_i = C_i^{equi}$ the A value vanishes.

Entropy Change in Open Systems. It may seem at first glance that the application of the classical expression of the second law of thermodynamics to biological systems leads to a paradoxical conclusion that the processes of life activity violate the thermodynamic principles. Indeed, complication and enhancement of the rate of ordering of organisms during their growth proceed spontaneously. But as follows from the second law, such spontaneous processes should be accompanied by a decrease rather than an increase in entropy. However it is evident that only in isolated systems the entropy increase in irreversible spontaneous processes should take place, whereas biological systems are open. Therefore the problem is to perceive in what way the entropy change is connected with the parameters of the process in an open system and to clarify whether it is possible to predict the general direction of irreversible processes in such a system by evaluating the change of its entropy. The main difficulty in solving this problem is that one should take into account the alteration of all thermodynamic values in time just when the processes proceed in the system. It is postulated that in an open system the overall entropy change dS may happen independently, either due to the exchange processes with the external environment d_eS, or as a result of internal irreversible processes (d_iS).

$$dS = d_eS + d_iS. \tag{6.6}$$

In all actual cases $d_iS > 0$, and only if the internal processes proceed reversibly and at equilibrium, then $d_iS = 0$. For isolated systems $d_eS = 0$, and so we have the classical expression of the second law:

$$dS = d_iS \geq 0.$$

Cell metabolism always has two groups of such processes, for example, the glucose intake from outside, release of its oxidation products outwards (d_eS) and oxidation of glucose in respiration processes (d_iS).

In photosynthesis, the inflow of free energy of light causes both the formation of a complex glucose molecule from simple compounds H_2O and CO_2 and reduction of the cell entropy $d_eS < 0$, while decomposition of glucose increases its entropy $d_iS > 0$. In compliance with the proportion of rates of changes in d_eS and d_iS the overall entropy dS in an open system may either increase or decrease with time. Let us differentiate expression (6.6).

$$\frac{dS}{dt} = \frac{d_eS}{dt} + \frac{d_iS}{dt}. \tag{6.7}$$

If $\frac{d_eS}{dt} < 0$ and $\left|\frac{d_eS}{dt}\right| > \frac{d_iS}{dt}$, the entropy of an open system will decrease $\frac{dS}{dt} < 0$. The negative value $\frac{d_e\ S}{dt} < 0$ or the inflow of negative entropy corresponds to the outflow of positive entropy into the external environment and the inflow of nutrients from the outside accompanied by the release of the free energy from the nutrients in the organism. But if $\left|\frac{d_eS}{dt}\right| < \frac{d_iS}{dt}$, then $\frac{dS}{dt} > 0$, which corresponds to the general degradation and decomposition of the system. In a stationary state $\frac{d_eS}{dt} < 0$, $\left|\frac{d_eS}{dt}\right| = \frac{d_iS}{dt}$ and $\frac{dS}{dt} = 0$.

Relationship between d_iS and Open System Parameters. Let us accept that the exchange processes of an open system with the environment proceed at equilibrium, and the only reason for the irreversibility and entropy increase in the system are its internal processes. In this case

$$\frac{d_iS}{dt} = -\frac{1}{T}\left(\frac{dG}{dt}\right)_{T,p}. \tag{6.8}$$

It can be shown that the rate of positive entropy formation within an open system depends on chemical affinity A and reaction rate ν

$$\frac{d_iS}{dt} = -\frac{1}{T}A\nu > 0. \tag{6.9}$$

Given $A - 0$ and $\nu = 0$, equilibrium is formed in the system and we have

$$\frac{d_iS}{dt} = 0. \tag{6.10}$$

Expression (6.9) has a simple meaning. It shows that the rate of the positive entropy formation in the system in irreversible chemical processes is directly proportional to its driving force A and rateν. It is obvious that the $\frac{d_iS}{dt}$ value is a variable because in the course of the chemical reaction the unstable concentrations of the reactants are always changing and hence values A and ν dependent on them change accordingly. Below we will analyze how value $\frac{d_iS}{dt}$ behaves at a stationary state where all concentrations become constant. When several chemical reactions proceed simultaneously in the system (j = 1, 2, …, n), the overall rate of positive entropy generation is dependent on their driving forces and rates,

$$T\frac{d_iS}{dt} = A_1\nu_1 + A_1\nu_2 + \ldots + A_n\nu_n = \sum_j A_j\nu_j > 0. \tag{6.11}$$

Coupling of Processes. Let us have two irreversible reactions (A_1v_1 and A_2v_2). Then

$$T\frac{d_iS}{dt} = A_1v_1 + A_2v_2 > 0. \tag{6.12}$$

The positive $T\frac{d_iS}{dt}$ value in (6.12) is provided when

$$A_1v_1 > 0, \quad A_2v_2 > 0 \tag{6.13}$$

or when

$$A_1v_1 < 0, \quad A_2v_2 > 0, \quad A_2v_2 > |A_1v_1|. \tag{6.14}$$

In this case the both reactions may be coupled. The first (coupled) reaction proceeds contrary to the difference of chemical potentials of its reagents (A_1 and v_1 have opposite signs) at the expense of the second (coupling) reaction. The energy released in the coupling reaction is not dissipated into heat, but is spent to maintain a coupled process proceeding with the increase in the chemical potentials of its initial products, i.e. with the increase in the free energy. Conditions (6.13) and (6.14) permit us to find the upper limit of the coupled reaction rate

$$v_1 \leq \frac{A_2v_2}{A_1}, \tag{6.15}$$

that is to relate the thermodynamic value of chemical affinity to the kinetic value of the reaction rate.

Onsager Relations. The driving forces and rates (inflows) should obviously be interrelated when the enhancement (reduction) of the driving force causes a corresponding increase (decrease) in the process rate. This concerns not only chemical reactions, but other irreversible processes as well. For example, the processes of heat transfer and substance diffusion across the membrane from one phase to the other involve driving forces (temperature and concentration gradients), and the fluxes denote the heat or substance transfer between the two phases. In all such cases the entropy increase can be written like this

$$T\frac{d_iS}{dt} = XI > 0, \tag{6.16}$$

where X is the generalized driving force, and I is the corresponding flux value.

If the system is near equilibrium then the driving forces and fluxes which are very small, are directly proportional to each other

$$I = LX \tag{6.17}$$

where L is the constant linear coefficient.

Take a chemical reversible reaction

$$C_1 \underset{\overleftarrow{v}}{\overset{\overrightarrow{v}}{\rightleftarrows}} C_2$$

where the value of the total rate (or flux) equals the difference between the rates of direct $\vec{v} = \vec{k}\,C_1$ and reverse $\overleftarrow{v} = \overleftarrow{k}C_2$ reactions:

$$I = \vec{v} - \overleftarrow{v} = \vec{k}C_1 - \overleftarrow{k}C_2.$$

Obviously at equilibrium $\vec{v} = \overleftarrow{v}$ and $I = 0$, and near equilibrium $I \cong 0$. The value of chemical affinity A near equilibrium is also very low

$$A = \mu_{C_1} - \mu_{C_2} << \; RT.$$

In this case I and A are proportional as well. If in an open system near equilibrium several processes occur at the same time, their thermodynamic proportions reflect their mutual effects. For two processes (I_1, X_1) and (I_2, X_2) these proportions are as follows

$$I_1 = L_{11}\,X_1 + L_{12}X_2$$

$$I_2 = L_{21}\,X_1 + L_{22}X_2 \tag{6.18}$$

where constant coefficients L_{11} and L_{22} show the dependence of a flux on its force, and coefficients L_{12} and L_{21} correspond to the mutual effect of the force of one process caused on the flux of the other process. They are called Onsager reciprocity coefficients and equations (6.18) are defined as Onsager linear relations. Near equilibrium we have

$$L_{12} = L_{21}.$$

Now we can determine the quantitative relationship between processes occurring concurrently in the cell without taking into consideration their molecular mechanisms.

Let us consider an active transfer of a substance across the membrane that occurs at the expense of the energy of the coupling metabolic process and therefore it can proceed against the concentration gradient of the transferred substance. Then

$$I_1 = L_{11} X_1 + L_{12} X_2 \text{ , and}$$

$$I_2 = L_{21} X_1 + L_{22} X_2 \text{ , } L_{12} = L_{21},$$

where the process (I_1, X_1) of coupled transfer moves against the force gradient X_1 (I_1, $X_1 < 0$) due to the energy of the coupling process (I_2, $X_2 > 0$). If no coupling takes place, then $L_{12} = L_{21} = 0$ and the processes occur independent of each other under the action of only "their own" driving forces

$$I_1 = L_{11} X_1 \text{ , } I_2 = L_{22} X_2.$$

The following value is introduced as a coupling measure

$$q = \frac{L_{12}}{\sqrt{L_{11} \; L_{22}}}.$$

It is $q = 0$ when the coupling is absent ($L_{12} = 0$), and $q = 1$ at completely coupled processes. At the initial moment of "triggering" the system, the high rate of the coupling process I_2 drops to minimal values and at the same time the X_1 value rises. On account of this, a stationary state is formed when the resultant coupled flux vanishes ($I_1 = 0$). If the system is entirely coupled, stationary state $I_2 = 0$ is formed for the coupling flux too. In this case, the system has no noticeable changes and the total energy of the coupling flux is spent on maintaining the force X_1. Let us imagine a turbine wheel immersed in a water stream. Depending on the depth of immersion, the number of the turbine wheel blades in water (X_1), velocities of the turbine rotation (I_1) and water flux (I_2) would change. These examples are valid not only for an active transfer, but also for other cases. So, in the system of respiration control in mitochondria, the substrate oxidation rate (I_2) depends on the ADP/ATP ratio, i.e. on the driving force X_1. In the state of mitochondria when the ADP concentration equals zero and no discernible formation of ATP occurs ($I_1 = 0$), the total energy is spent on maintaining the maximal phosphate potential (X_1^{max}). An addition of uncouplers results in a decrease in X_1, but when $I_1 \neq 0$ it causes the acceleration of the coupling flux I_2.

The coefficient of energy transformation in coupling processes is I_1X_1/I_2X_2 and in mitochondria its value can make 80-90%. The use of Onsager equations permits obtaining characteristics of macromolecular complexes, i.e. biological energy transformers, without a detailed analysis of molecular mechanisms of their functioning.

Prigogine Theorem. It has already been shown that at a stationary state of an open system

$$\frac{dS}{dt} = \frac{d_eS}{dt} + \frac{d_iS}{dt} = 0\,,$$

$\frac{d_eS}{dt}$ and $\frac{d_iS}{dt}$ members differing from zero. The question arises whether it is possible to predict the formation of a stationary state in an open system by analyzing the change in $\frac{d_iS}{dt}$ with time. Let us have two processes occurring concurrently in an open system near equilibrium, for which Onsager relations (6.18) and (6.19) are valid. When one of the processes (I_1X_1) reaches a stationary regime, then for it $I_1 = 0$. Determine the following value

$$T\frac{d_1S}{dt} = I_1X_1 + I_2X_2 = L_{11}X_1^2 + 2L_{11}X_1X_2 + L_{22}X_2^2 > 0, \quad (6.19)$$

which has a positive quadratic form. When approaching the stationary state, the driving force X_1 and flux I_1 change in a certain way so that eventually $\overline{I} = 0$. Let us see how $T\frac{d_1S}{dt}$ depends on the change of X_1. To this end we take partial derivative $\frac{\partial\left(T\frac{d_1S}{dt}\right)}{dX_1}$ at constant X_2 and T. From (6.20) we get that

$$\left[\frac{\partial\left(T\frac{d_1S}{dt}\right)}{dX_1}\right]_{X_2=\text{const}} = 2(L_{11}X_1 + L_{12}X_2) = 2I_1.$$

But at a stationary state, $I_1 = \overline{I}_1 = 0$. Consequently, the equations

$$\left[\frac{\partial\left(T\dfrac{d_1S}{dt}\right)}{dX_1}\right]_{X_2=\mathrm{const}}=0 \qquad \text{and } I_1=\overline{I}_1=0$$

are equivalent. The vanishing of the partial derivative of $T\left(\frac{d_1S}{dt}\right)$ with respect to X_1 in a stationary point $I_1 = \overline{I}_1 = 0$ shows that at the stationary state the positive function $\frac{d_iS}{dt}$ has the extremum and accordingly is positive. So, while reaching the stationary state, the rate of entropy formation within the open system is decreasing monotonously, gradually coming close to its minimal constant positive value. This is the criterion of the direction of irreversible processes in open systems that proceed near equilibrium where Onsager relations are effective. Provided the system has a stationary state where value $\mathrm{T}\frac{d_iS}{dt}$ is minimal, any deviation from the stationary point caused by disturbances would lead to its increase. But then in view of the Prigogine theorem, the values of forces and fluxes in the system should change so that the rate of entropy formation $\mathrm{T}\frac{d_iS}{dt}$ would again change, and the system would return to the stationary point. This result illustrates the stability of the stationary state in question.

As follows from the monotonic character of $\mathrm{T}\frac{d_iS}{dt}$ changes near equilibrium the stationary state cannot represent an autooscillation regime. Indeed, in this case variable concentrations in the system and as a result, values I and X change regularly which is incompatible with the unidirectional monotonic change of $\mathrm{T}\frac{d_iS}{dt}$ and its constancy in the stationary point.

Experimental measuring of the rate of entropy formation in the system can be performed using calorimeters by studying heat fluxes accompanying the entropy formation upon irreversible changes in the system. It was demonstrated in experiments on biological objects that, for example, both the rate of heat generation and respiration decrease continuously during embryo development beginning from the initial stages of organism development and attain constant values at the stationary phase of growth. However it should be noted that the level of thermogenesis can alter in the course of organism development not only due to changes in the values of driving forces and fluxes. The heat generation in organisms depends also on the state of membrane structures and the level of coupling of the oxidative phosphorylation processes. At length, essential is the fact that biological systems are far from equilibrium where the proportionality of I and X (6.18) or reciprocal relations (6.19) are broken. This is of special importance for biochemical processes where the most common transitions are accompanied by changes in ΔG by 1-2 kcal/mol, while the Onsager relations are effective at $\Delta G \leq 0.2$ kcal/mol. Under such conditions when stationary states are far from equilibrium, the Prigogine theorem generally is not valid (an autooscillation regime).

Thermodynamics of Active Transport. Let us analyze the thermodynamics of active transport of one (sodium) ion not associated with the transfer of other substances. For the sake of simplicity of argumentation, accept that it is possible to indicate the metabolic process as a driving force of the active transport.

Let us denote the rate of the cation active transport as I_+^a and the rate of the driving process in metabolism as I_r then

$$I_+^a = L_+^a X_+ + L_{+r}^a A,$$
$$I_r = L_{+r}^a X_+ + L_r^a A, \quad (6.20)$$

where X_+ is the negative difference of electrochemical cation potentials (moving against "its own" driving force gradient) and A is the affinity of the metabolic reaction necessary for the transport. In the case of one metabolic reaction (ATP hydrolysis), the rates of intake and generation of all metabolites are associated stoichiometrically. Therefore to estimate the metabolic rate one may take, e.g. the rate of O_2 consumption upon respiration. Then the affinity A may be expressed as a negative change in the entire thermodynamic potential of the metabolic reaction (ATP hydrolysis) per mole of O_2 consumed.

It should be reminded that the electrochemical potential of ion $\bar{\mu}$ is the sum of its chemical potential μ and electric potential φ on the membrane

$$\bar{\mu} = \mu_0 + RT1nC_+ + ZF\varphi,$$

where F is the Faraday number (96,500 coulomb/mol), Z is the ionic valency, and C_+ is the ion concentration. Phenomenological coefficients L^a_r link the active transport ($I_+{}^a$) and metabolism (I_r) because of the coupling and interference of the fluxes and forces of these processes. So, for a monovalent ion, the driving force X_+ is determined as

$$X_+ \Delta\bar{\mu} = RT\ln\frac{C^i_+}{C^0_+} + F\Delta I. \tag{6.21}$$

By placing the same solutions ($\Delta C = C_+^{\,i} = C_+^{\,o} = 0$) on both sides of the membrane and by varying the Δφ value, one can determine phenomenological coefficients from (6.20)

$$L^a_{\ +} = \frac{\partial I^a_+}{\partial X^+} = -\frac{\partial I^a_+}{\partial (F\Delta\phi)} = -\frac{\Delta I^a_+}{\Delta (F\Delta\phi)} \, (\Delta C = 0,\ A = \text{const}),$$

$$L^a_{\ +r} = \frac{\partial I_r}{\partial X^+} = -\frac{\partial I_r}{\partial (F\Delta\phi)} = -\frac{\Delta I_r}{\partial (F\Delta\phi)} = -\frac{\Delta I}{\partial (F\Delta\phi)}$$
$$\left(\Delta C = 0,\ A = \text{const}\right).$$

Coefficients L^a_+ and $L^a_{\ +r}$ are found from the slope of corresponding straight lines $\Delta I^a_{\ +} = -L^a_{\ +}\Delta\left(F\Delta\varphi\right)$ and $\Delta I_r = -L^a_{\ +r}\,\Delta\left(F\Delta\varphi\right)$. The constant A values at transitory changes in X_+ facilitate a successful thermodynamic analysis.

Thus, for a sodium flux in frog skin tissues, the equations are as follows

$$\begin{aligned} I^a_{\ \text{Na}} &= L_{\text{Na}} X_{\text{Na}} + L_{\text{Na},r} A, \\ I_r^{\ \text{sb}} &= L_{\text{Na},r} X_{\text{Na}} + L_r A. \end{aligned} \tag{6.22}$$

The $I^a_{\ \text{Na}}$ value is accepted to be positive when the flux moves from the external (mucous membrane) to the internal (serous membrane) surfaces of the tissue. The $I_r^{\ \text{sb}}$ value is the rate of that portion of the total respiration process (oxygen consumption) which is directly connected with the transport and exceeds the level of basal metabolism of substances. It denotes the suprabasal flux of the metabolic reaction. Obviously, it is critical to preserve the system parameters and be able to purposefully vary X_{Na} at constant A in experimental conditions in order to determine the coefficients and maintain linear dependencies in equations (6.22). The experiments were carried out on frog skin with varying the X_{Na} values caused by Δφ changing. At the same time the composition of the washing liquid and sodium concentration remained unchanged. In such conditions $X_+ \cong -F\Delta\varphi$ and the equations for transport look like this

$$I^a_{\mathrm{Na}} = \mathrm{L}_{\mathrm{Na}}(-F\Delta\varphi)+L_{\mathrm{Na},r}A,$$
$$I_r^{\mathrm{sb}} = \mathrm{L}_{\mathrm{Na,r}}(-F\Delta\varphi)+L_rA,$$

where $\Delta\varphi = \varphi^i - \varphi^0$.

It was proved in direct experiments that the rate of active transport on frog skin I^a_{Na} depends linearly on the $\Delta\varphi$ value, $\Delta\varphi$ being symmetrically changed in the range from 0 to ±80 mV. The I_2 value can be determined from the oxygen consumption using oxygen electrodes. It was found that at symmetrical disturbances of the potential the relation of I_r and $\Delta\varphi$ was linear in the range from 0 to ±70 mV. The dependency of I^a_{Na} and I_r^{sb} versus the external concentration of sodium in conditions of its constant internal concentration at constant zero difference of electric potentials ($\Delta\varphi = 0$) was studied as well. Under such conditions, it was also observed that the rates of active transport I_{Na} and suprabasal oxygen consumption I_r^{sb} depend linearly on the difference of chemical potentials $\Delta\bar{\mu}_{\mathrm{Na}}$ on the membrane. But if X_{Na} is changed by varying the internal concentration of sodium, no linearity is observed any longer. This is the result of changes in the microstructure and composition of the membrane *per se*.

Such experiments were successfully performed for studying the active transport of protons using equations of non-equilibrium thermodynamics for two fluxes. In all cases, the varying of X_+ permits estimating phenomenological coefficients and affinity A of the driving metabolic reaction. Analogous formalism has been effectively applied to describe the processes of phosphorylation in mitochondria and chloroplasts. It is generally accepted that these objects have tight coupling of the three main processes underlying the bioenergetics of cell membranes: electron transport with substrate oxidation (I_0, A_0), phosphorylation of ADP with ATP formation (I_{P}, A_{P}), and translocation of protons across the coupling membrane ($I_{\mathrm{H}}\Delta\bar{\mu}_{\mathrm{H}}$). Of the key importance is the transmembrane circulation of protons, which is induced by electron transfer and, in its turn, "triggers" the ATP synthesis. The phenomenological description of the system includes three corresponding equations

$$I_p = \mathrm{L}_pA_p + \mathrm{L}_{pH}\,\Delta\bar{\mu}_H + \mathrm{L}_{p0}A_{p0,}$$
$$I_{\mathrm{H}} = \mathrm{L}_{pH}A_p + \mathrm{L}_{\mathrm{H}}\,\Delta\bar{\mu}_{\mathrm{H}} + \mathrm{L}_{0\mathrm{H}}A_{0,}$$
$$I_0 = \mathrm{L}_{p0}A_p + \mathrm{L}_{0\mathrm{H}}\,\Delta\bar{\mu}_{\mathrm{H}} + \mathrm{L}_0A_{0,} \tag{6.25}$$

where the meaning of coefficients L is obvious. Equations (6.25) can be used when the affinity values A_0 and A_{P} may be varied within a wide range and the $\Delta\bar{\mu}_H$ value can be estimated from the difference of pH and electric potentials on the coupling membranes. It was determined that in this case there is also a linear dependence between the forces and fluxes, which should facilitate determining coefficients L experimentally. To achieve this, the experimental conditions can be simplified by maintaining A_0 = const without changing the $\Delta\bar{\mu}_H$ value and by achieving a stationary state when $I_H = 0$. The $\Delta\bar{\mu}_H$ value may be kept equal to zero. In both cases equations (6.25) will be simplified so that they can be used for the estimation of the required data. This approach is used as a starting point for discussing various hypotheses of energetic coupling. Particularly, it is presumed in the chemosmotic hypothesis that only proton translocation is directly associated with the ATP formation rather than with the electron transfer. In extreme case, coefficient I_{p0} should equal zero. Undoubtedly it should be clearly perceived that the thermodynamic analysis can be helpful in estimating the energetic efficiency and the level of coupling of processes, but it does not suggest anything about their actual molecular mechanisms.

7

Thermodynamics of Systems Far from Equilibrium

Methods of classical thermodynamics can be used for concluding whether an isolated system can perform a spontaneous transition from one state into the other. An open system has stationary states that can be either near or far from thermodynamic equilibrium. The problem of whether an open system is able to make transition from some initial state into the final stationary state can be solved by comparing the rates of entropy formation in these states, provided they both are within the range of linear thermodynamics, that is near thermodynamic equilibrium. However far from equilibrium it is impossible to make univalent conclusions on how the rate of entropy formation changes. The evolution of such non-equilibrium dynamic systems is determined primarily by both the kinetics of interactions of constituents and the movement of the system along phase trajectories rather than by statistical ordering of its initial and final states. Such systems have a limited number of final states and behave like "chemical machines". That is why propagation of thermodynamic ideas over non-equilibrium system behavior can provide only an additional characteristic of their stationary states far from equilibrium, while the position and modes of attaining such states depend mainly on kinetic equations.

Stability of Stationary Points. Let us briefly discuss the thermodynamic features of stability of stationary points, their properties being analyzed earlier in Chapters 1–4. Assume that the stationary state of the system was disturbed bringing about deviations in the values of forces and fluxes from their stationary values ($\overline{X}$ and I). It appears that if the initial stationary state is stable, the product of the "disturbance" values δI and δX must be positive

$$\delta I \cdot \delta X > 0.$$

This is the criterion of the stability of stationary states far from equilibrium. However, attempts to find general thermodynamic criteria for the motion to the stationary state far from equilibrium are not successful. The reason is the deterministic nature of behavior of kinetic systems where contrary to equilibrium systems, the idea of entropy is not decisive in predicting the direction of transition processes.

Compare the type of stationary point stability to its distance from thermodynamic equilibrium. Only stable stationary states of the "node" type are allowed near equilibrium. As the distance from the equilibrium increases the values X and I will also increase (Fig. 7.1) and the system may leave the range of linear thermodynamics without losing its overall stability. The "stable focus" point corresponds to this event. Though it is possible that upon moving away from equilibrium, the system will undergo a bifurcation change and instability will take place. Unstable stationary points that are inherent to the system far from equilibrium are "saddles" or "unstable focuses". At the bifurcation point, where stability is lost, the product $\delta I \cdot \delta X$ becomes negative

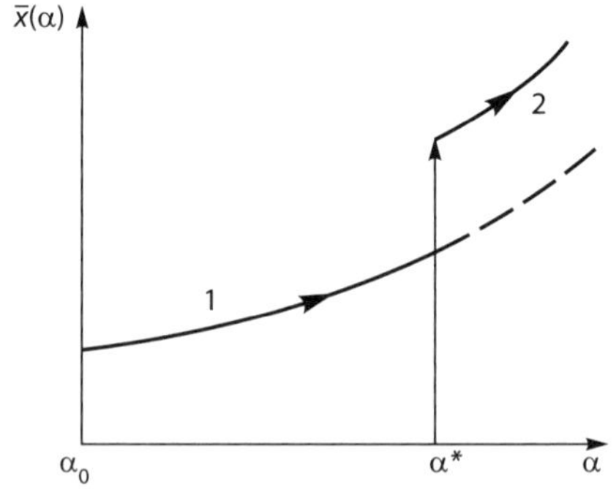

Figure 7.1 Dependence of the stationary concentration of a component element versus parameter α that is a measure of deviations.

($\delta I \cdot \delta X < 0$) which agrees with the thermodynamic threshold of instability origination in the system. In this case it is said that thermodynamic fluctuation appears leading the system away from the unstable point which may be a reason for the system collapse. However at definite parameter values this fluctuation seems to stimulate the system transfer to a new state to which the stability is passed over. For example, this may be a limit cycle near the unstable focus. The formation of dissipative structures in distributed systems is also preceded by the disturbance of thermodynamic stability far from equilibrium. Finally, trigger transitions between stable stationary states (see Fig. 3.4) occur at the boundary of stability on the curve of stationary states, when thermodynamic stability is broken, and the system makes an abrupt transition from one stable state to another.

So, thermodynamic features of stability of stationary states are compatible with corresponding mathematical features and can serve as their additional characteristic. But far from equilibrium, no general thermodynamic criteria exist for the direction of movement of the open system because its behavior is conditioned by its dynamic properties and kinetic regulatory mechanisms rather than by general statistical rules. This specificity underlies the complexity of entropy and information concepts when describing the general properties of biological systems.

Entropy and Information. In accordance with the Boltzmann formula, entropy is determined as the logarithm of the number of microstates possible in the given macroscopic system

$$S = k_B \ln W \tag{7.1}$$

where $k_B = 1.38 \cdot 10^{-16}$ erg·degr^{-1} or $3.31 \cdot 10^{-24}$ entropy units (1 e.u. = 1 cal·degr^{-1} = 4.1 J/K), or $1.36 \cdot 10^{-23}$ J/K is the Boltzmann constant, and W is the number of microstates (for example, the number of possible ways to distribute gas molecules in a vessel). Namely to this effect entropy is the measure of the system disorder and chaotization. Real systems possess stable and unstable degrees of freedom, for instance, related to solid walls of the vessel and the gas molecules in it. The notion of entropy is associated just with unsteady degrees of freedom when chaotization of a system may take place and the number of allowable microstates far exceeds unity. In an ideal totally stable system only a single solution is fulfilled, i.e. the number of ways to realize this particular macrostate of the system equals unity ($W = 1$) and the entropy equals zero. The notion of entropy as well as thermodynamic concepts can be used in biology with respect to certain metabolic processes rather than for describing at large the behavior and general biological properties of organisms. The relationship between entropy and information in the information theory was recognized for statistical degrees of freedom. Assume that we obtained information on what is an actual way of all the possible ones to realize a certain macrostate of the system. Apparently, the amount of the information obtained is larger the higher is the initial uncertainty or entropy of the system.

In keeping with the information theory, in this case the amount of information concerning a single actual state of the system will be as follows

$$I = \log_2 W. \tag{7.2}$$

A unity of amount of information (bit) is the information contained in a reliable message when the number of initial possible states is $W = 2$:

$$I = \log_2 2 = 1 \text{ bit}. \tag{7.3}$$

For example, the message on what side the coin fell down when thrown in the air contains 1 bit of information. A comparison of formulas (7.1) and (7.2) allows finding relationship of the entropy (in entropy units) and information (in bits)

$$S\,(\text{e.u.}) = 2.3 \cdot 10^{-24} = 1 \text{ bit}. \tag{7.4}$$

Now let us try to formally estimate the amount of information contained in the human body consisting of 10^{13} cells. Using formula (7.3) we get

$$I = \log_2 10^{13} \sim 10^{13} \log_2 10^{13} \sim 4 \cdot 10^{14} \text{ bits}.$$

This amount of information would be required initially to realize the only possible accurate location of cells in the organism. It is equivalent to a quite insignificant decrease in the entropy of the system

$$\Delta S = 2.3 \cdot 10^{-24} \cdot 4 \cdot 10^{+14} \sim 10^{-9} \text{ e.u.} \sim 4 \cdot 10^{-9} \text{ J/K}.$$

If it is accepted that the human organism has also a unique character of the sequence of amino acid residues in proteins and nucleotide residues in the DNA, the overall amount of information contained in the human body will be

$$I \sim 1{,}3 \cdot 10^{26} \text{ bits},$$

which is equivalent to a small reduction of entropy by $\Delta S \sim 300$ e.u. = 1200 J/K. In metabolic processes, this decrease of entropy is readily compensated by its increase upon oxidation of 900 glucose molecules. Thus, a comparison of formulas (7.1) and (7.2) demonstrates that biological systems do not possess formally any increased information capacity as compared to other non-living systems consisting of the same number of structural elements. At first glance, this conclusion conflicts with the significance of the role of informational processes in biology.

But the relation of I and S values in (7.4) is valid only with respect to the information on which of the total number of microstates W is realized at the moment. This microinformation associated with the arrangement of all atoms in the system cannot be recorded and stored because any of such microstates would rapidly transfer to another one due to thermal fluctuations. The real value of biological information is determined not by its amount, but first of all by the possibility to record, store, process, and further transmit the information to be used in the organism life activity.

The basic condition of perception and storage of information is the ability of the receptor system due to the obtained information to switch to one of the stable states *a priori* inherent to the system because of its organization. That is why informational processes in organized systems are associated only with definite degrees of freedom. The storage of information must proceed with a concurrent energy loss in the receptor system so that it could be stored for a sufficient time and not lost due to thermal fluctuations. It is here that microinformation, which the system could not store, turns into macroinformation, which the system stores and then can pass it to other acceptor systems. Entropy is recognized to be the measure of the diversity of microstates that cannot be memorized by the system, while macroinformation is the measure of the diversity of their states, which should be remembered by the system.

For instance, the information capacity of DNA is determined only by the number of specific nucleotides and not by the total number of microstates including vibrations of all atoms in the DNA chain. The process of information storage in DNA consists in fixation of specific positions of nucleotides that are stable because of the chemical links formed in the chain. Further transmission of the genetic information occurs as a result of biochemical processes in which energy dissipation and formation of consistent steady chemical structures provides for the efficiency of biological information processing. On the whole, information processes are very important in biology. On the molecular level, they happen not only upon storage and processing of the genetic information, but also upon mutual recognition of macromolecules, provide for the specificity and direction of enzyme reactions, and are of importance for interaction of cellular membranes and surfaces. Physiological receptor processes, playing an independent information role in the organism life activity, are also based on interactions of macromolecules. In all cases, macroinformation is generated initially as conformational changes coupled to energy dissipation on definite degrees of freedom in the interacting macromolecules. As a consequence, macroinformation is recorded as a set of energetically rather deep conformational substates, which permit storing this information for a time period required for its further processing. The biological importance of this macroinformation is realized with respect to the biological organization and cell structures, where further processes take place causing corresponding physiological biochemical effects.

It can be stated that living systems regulate directionally biochemical reactions at the level of single macromolecules, which in general define macroscopic properties of biological systems.

Such properties are not inherent even to the latest state-of-the-art devices such as submicron processors in which electron fluxes are monitored with unavoidable energy losses. Below it will be shown that in biomembranes, electron fluxes are regulated with respect to the transfer of every individual electron along the chain of macromolecular carriers.

Moreover it will be demonstrated that energy transformation in biological processes occurs in macromolecular energy-transforming nanosized "machines".

Small dimensions specify low values of energy gradients and, as a result, approximate the operation of such machines to the conditions of thermodynamic reversibility. As known, this increases the energetic efficiency (coefficient of efficiency) of energy transformation. Such nanosized molecular machines optimally combine the maximal energy yield and the low level of energy dissipation corresponding to the low rate of entropy generation in the system.

Small differences in the values of redox potentials of separate carriers in the chain of photosynthesis and respiration illustrate the above statement providing for conditions close to the reversibility of individual processes in electron transport.

The analysis of the operation of molecular motors coupled to energy transformation challenges the development of thermodynamics of small-scale systems, in which the values of energy differences at elementary stages of working cycles are comparable to those of thermal fluctuations. Indeed, the average value for the overall energy of a macrosystem (ideal gas) consisting of N particles and distributed over them in accord with the Gauss law is $3/2N\ k_BT$. The size of random fluctuations of this value is about $1/\sqrt{N}$ and is insignificant with respect to the average value for a system consisting of a large number of particles. However at low N, the size of fluctuations approaches the average energy value for such a small system that can be only of several k_BT units.

For example, a kinesin molecule smaller than 100 nm moves along microtubules carrying cell organelles and making 8-nm steps every 10-15 milliseconds due to the energy of ATP hydrolysis (20 k_BT). At every step the "kinesin motor" makes work of 12 k_BT with the coefficient of efficiency 60%. In this sense, kinesin is one of various molecular machines making use of the energy of phosphate bond hydrolysis in different processes including replication, transcription, repairing and the like. The small size of such machines can be helpful in absorbing the energy of large thermal fluctuations from the environmental space. Undoubtedly, on average when the molecular motor moves along its dynamic trajectory, the performance of work results in thermal energy release. But at separate stages of the working cycle, the accidentally absorbed energy of thermal fluctuations combined with the "directed" energy of hydrolysis of phosphate bonds may contribute to the relation of the free energy change and the work performed. In this case, thermal fluctuations can cause already evident deviations from the averaged dynamic trajectories. Consequently, such small systems cannot be adequately described only in terms of classical thermodynamics. At present these problems are being actively worked on together with the progress in nanotechnologies associated with the development of nanosized molecular machines.

It should be noted once again that biochemical processes of energy transformation, in which useful chemical work is performed, are per se only suppliers of elements for self-organization of biological structures and thus for generation of information in biological structures.

It is precisely to biochemical reactions that the basic principles of chemical thermodynamics and, in particular, the fundamental concept of chemical potential, as a measure of the dependence of the number of possible microstates on the number of particles in the system, are applicable.

In fact, a chemical reaction is a result of redistribution of the number of moles or the number of reagent molecules provided the total number of atoms in the reaction is unchanged. These redistributions are associated with the splitting and formation of chemical bonds between atoms in the molecules coupled to thermal effects. It is in the range of linear thermodynamics that their general direction obeys the Prigogine theorem. Figuratively speaking, a biochemical reaction generates structural elements and delivers them to the site of self-assembly of stable "informational" macromolecular complexes, i.e. carriers of information. The self-assembly itself occurs spontaneously and naturally proceeds with the general free energy decrease: $\Delta F = \Delta U - T\Delta S < 0$. Actually, upon formation of a stable ordered structure, the absolute value of the internal energy of formed structures ΔU should be larger than the decrease in the entropy ($-T\Delta S$), $|\Delta U| > |T\Delta S|$, so that the free energy decreases $\Delta F < 0$.

Let us remind that during prebiotic evolution, stable structural "bricks" of living matter (amino acids, nucleotides, sugars) were formed spontaneously, abiogenically from inorganic simple compounds, without living systems, at the expense of external sources of energy (light, electric charges) required for overcoming the activation barriers of reactions of synthesis.

Generally, the direct generation of biological information on the macromolecular level leads to a decrease in the structural entropy (generation of negative entropy). This entropy reduction is compensated by the formation of stable associations in the generated information structures. However, the balance of "thermodynamic" entropy in an open system is determined by the relation of the driving forces and fluxes in a group of chemical processes that create conditions for synthesis of information structures.

Apparently, the calculation of the overall balance between structural and thermodynamic entropy in a living system has a purely arithmetic meaning. It is conditioned by two interrelated groups of processes, however differing in their nature, with no direct compensation for the entropy change between them.

8

Physicochemical Principles of Biopolymer Structure

A macromolecule as the basic structural unit of living matter has a great number of atoms and atomic groups. Their thermal motions, turns and rotations around single bonds specify the diversity of intramolecular degrees of freedom which adds statistical properties to the macromolecule. On the other hand, atoms in the same macromolecule are linked by chemical bonds, long- and short-range interactions that makes conformational rearrangements quite deterministic. So, a biological macromolecule has peculiar properties based on close interaction of statistical and deterministic (mechanical) degrees of freedom. In simple chemical processes, the reaction product is generated in solution as a result of active collisions of reagent molecules. In contrast to this, the result of macromolecule functioning in biochemical processes is attained via interactions of the parts of the whole active macromolecular complex. In solutions, the temperature growth causes an increase in the number of active collisions between molecules, whereas in macromolecular complexes the same factor may affect their structural organization and thus the mechanism and efficiency of intramolecular interactions. Strictly speaking, for such systems it is impossible to apply the concept of chemical potential as the driving force of the process that depends just on the number of reagent molecules. In case of macromolecular complexes, the reaction is not determined by their number as such, but by intramolecular interactions in each of them. This is well demonstrated, e.g., in enzyme catalysis.

The primary task of molecular biophysics is to reveal the nature of intramolecular dynamics proceeding from interactions of atomic groups that determine macromolecular conformation. On this base we will pass to the consideration of electron properties and physical principles of functioning of macromolecules (proteins) in biochemical processes.

Coil and Globule. The polymer chain in which only neighboring links interact with each other, folds into a coil having a large number of conformations, transitions between of the latter being due to micro-Brownian motion of the chain parts. Such a coil has no fixed internal conformation; it appears as though it "breathes" all the time, the amplitude of a "sigh" being close to the coil size. The relative position of separate parts of the coil obeys utterly the statistical rules. But if atoms distant from each other interact volumetrically, the total pattern basically changes. In actual macromolecules, volumetric interactions of chain links create an internal field, which affects the state of a globule to form a tightly packed core. In difference to the coil, the globule has a specific three-dimensional structure. The core of the large globule is spatially homogeneous where a constant concentration of links is higher compared to that of the edge of the globule (Fig. 8.1). Temperature transitions between the states of the coil and globule at the same time are transitions between different phases accompanied by changes in the aggregation state of the macromolecule. These processes in biomacromolecules have been studied in more detail in proteins. Protein globules experience order–disorder transitions in a relatively small temperature range and in this respect resemble phase transitions of the first order. In experimental studies with the use of calorimeters, heat is usually supplied to the protein sample at a constant rate with concurrent

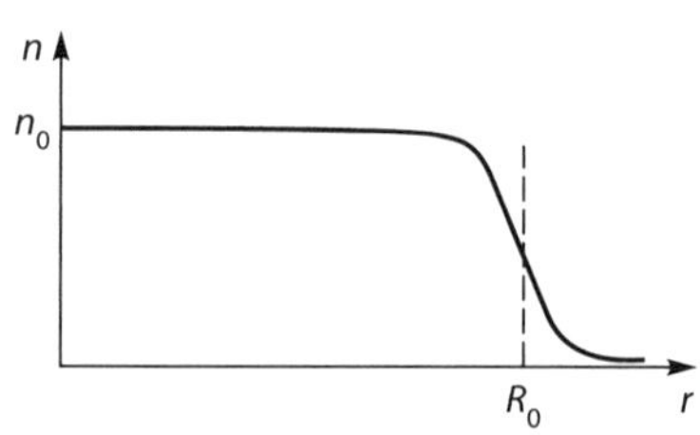

Figure 8.1 Distribution of the density of links in globule depending on the distance from the center of the globule. R_0 is the radius of the globule, and n_0 is the number of links.

recording of the rate of an increase in the protein temperature. Thereby it is possible to determine the protein heat capacity at different temperatures. It has been found that thermal denaturation of protein and order–disorder transitions go along with a simultaneous change of its heat capacity ($\Delta C_p \sim 0.2$–0.6 J·g^{-1}·K^{-1}). These changes cannot be caused only by thermal excitation of intramolecular degrees of freedom, but they are evidence of structural rearrangements in the macromolecule itself. The detailed character of such rearrangements in the protein globule can be understood taking into account specific mechanisms of volumetric interactions in protein.

Types of Volumetric Interactions. The primary structure of a polymer chain is determined by chemical and valence interactions. Volumetric interactions determine predominantly the secondary structure of macromolecules. The universal criterion of stability of molecular structure is the minimum on curve $U(r)$ for the dependence of energy interaction versus the distance between the interacting parts. Figure 8.2 shows curve $U(r)$ with minimum $r = r_0$ in the case of two particles (diatomic molecule). At small distances repulsive forces prevail, while at large distances attraction predominates. Given $r = r_0$, forces of attraction and repulsion balance each other. The energy value $U(r)$ of free particles at $r \to \infty$ vanishes, and the energy of the stable structure formed by the particles is negative $U(r_0) < 0$. At small distances, where the particles repulse, this energy is positive $U(r) > 0$. The minimum of $U(r_0)$ complies with the interaction energy maximal by its absolute value and negative by the sign. Van-der-Waals forces play a significant role in the formation of protein secondary structure. They have electromagnetic nature and are related to the interaction of electric dipoles in neighboring molecules. The most prevalent are dispersive interactions

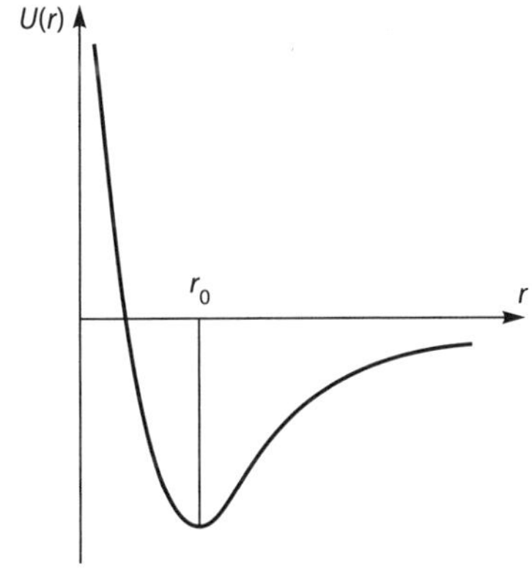

Figure 8.2 Electronic term $U(r)$ for a diatom molecule.

of molecules that have no constant dipole moments. These forces are quanto-mechanical by their nature. Motion of an electron as a quantum particle cannot be described with an exact mechanical trajectory but is rather of a smeared character. Electrons do not possess strictly defined values of coordinate (x) and impulse (p) ($p = mv$, i.e. the product of mass m by velocity v).

Uncertainties in the values of coordinate Δx and impulse Δp are related by the quantum-mechanical relation of uncertainties

$$\Delta x \cdot \Delta p \cong h. \tag{8.1}$$

This means that even in the ground unexcited state fast displacements of an electron charge from the equilibrium position take place, so that "instantaneous" dipole moments are generated in the molecule. The emergence of such dipole moments in one molecule induces its emergence in the adjacent molecule. The two fast-changing dipole moments are associated and mutually attracted through their interaction. The energy of attraction of two instantaneous dipoles, or the energy of dispersive interaction, rapidly diminishes with increasing distance

$$U_{\text{disp}} \sim \frac{1}{r^6}. \tag{8.2}$$

In addition to dispersive interaction, electrostatic attraction of constant dipoles in polar molecules is also possible. Besides, induction interactions may occur between the constant dipole moment in one molecule and the dipole induced by it in the adjacent polarized molecule. The total van-der-Waals interaction of two molecules is dependent on the contribution of all types of dipole interactions and makes from 1.0 kcal/mol up to several tens of kcal/mol. For many biological macromolecules, the depth of the energy minimum formed because of the van-der-Waals attraction is 1-3 kcal/mol, the order of this value is comparable to the value of thermal energy (0.6 kcal/mol) at room temperature ($RT = 2$ cal·K^{-1}·mol^{-1}·300, K = 0.6 kcal·mol^{-1}).

Not only the attraction ($U_{\text{attr}}(r) \sim \frac{1}{r^6}$), but the repulsion at close distances ($U_{\text{rep}}(r) \sim \frac{1}{r^{12}}$) should be taken into consideration in the expression for the total energy or total potential.

Summation of these values gives

$$U_{i,k}(r) = \frac{A}{r_{i,k}^6} + \frac{B}{r_{i,k}^{12}}, \tag{8.3}$$

where *A* and *B* are constants of attraction and repulsion, and $r_{i,k}$ is the distance between interacting atoms (*i* and *k*). At given values of *A* and *B* it is possible to find the minimal distance r_0 or the position of minimum $U(r_0)$ at which the system of two different atoms or small atomic groups (CH_3 and CH_2) is stable. So, for pairs of carbon atoms the minimal distance is 3.0 E and for H…H it is 2.0 E.

Along with van-der-Waals forces, a great role in stabilization of biostructures belongs to hydrogen bonds and electrostatic interactions of charged and polar groups. For example, hydrogen bonds stabilize the secondary structure of polypeptide chains. Electrostatic interactions, attraction and repulsion (8.3) and also the energy of delocalization of electrons contribute to the hydrogen bond energy. The values of hydrogen bond energy vary greatly (3–8 kcal/mol). Thus, hydrogen bond O–H…O has the energy of 8.6 kcal/mol.

Electrostatic interactions are assigned by the formula

$$U_{\text{el.stat}} = -\sum \frac{q_i q_k}{\varepsilon r_{ik}}, \tag{8.4}$$

where q_i and q_k are charges of atoms *i* and *k*, $r_{i,k}$ is the distance between the atoms, and ε is the dielectric constant ($\varepsilon \sim 3.5$ for proteins).

Internal Rotation and Rotational Isomerism are of great importance for the conformational structure of macromolecules. The energy of short-range interactions of atomic groups depends on the distances between them, which in their turn change upon rotation of these groups around single bonds. When closely positioned, atoms without valence bonds start repulsing so that the energy potential retarding the rotation of atomic groups originates. The energy of rotation of atomic groups around single bonds makes the major contribution to the total conformational energy of the polymer chain.

It is seen in Fig. 8.3 that an ethane molecule has minimum of the conformational energy in the trans-conformation and maximum in the cys-conformation. The value of the energy barrier, or the retarding potential, is almost 3 kcal/mol, when one trans-conformation turns into another via a cys-form upon rotation around the C–C bond by 120 degrees. The dependence of the potential of internal rotation on the rotation angle φ is determined using the expression

$$U(\varphi) = \frac{U_0}{2}(1 - \cos 3\varphi), \tag{8.5}$$

where U_0 is the barrier height.

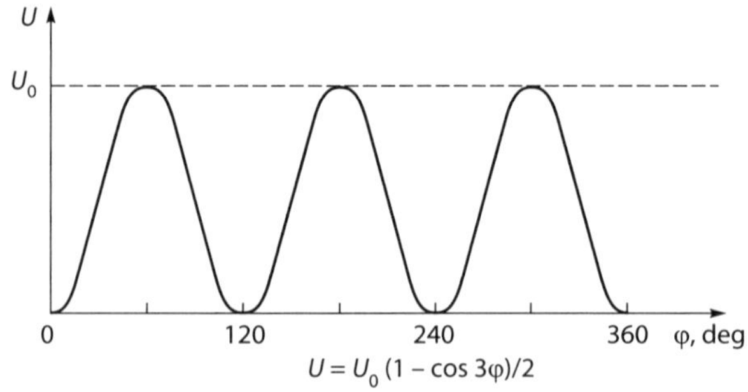

Figure 8.3 Scheme of the dependence of potential energy of internal rotation in ethane versus the rotation angle. Angles of 0°, 120° and so on correspond to trans-conformations, and 60°, 180° and so on to cys-confomations.

The overall conformational energy of polymer depends on complementary angles of reciprocal rotation of links around single bonds. Such a system where the energy of constituent elements is dependent on their interaction with each other is called a cooperative system.

Conformational Energy of Polypeptide Chain is determined by all types of volumetric interactions and depends on the energy of internal rotation of side chains of amino acid residues around single bonds. The common structure of a polypeptide chain is shown in Fig. 8.4. As it seems at first glance, side groups can rotate around all single bonds: N_i – C^{α}_i by angle φ_I, $C^{\alpha}_iC_i$ by angle ψ_I, and C_i – N_{i+1} by angle ω_i. In this case, the interference of the links upon changing their position can be transmitted along the chain via a great number of bonds, and then it will be difficult to take into account their contribution into the overall conformational energy. But actually the dual nature of peptide bond C^{α}_iO–$N_{i+1}H$ prevents rotation around it. It is specified by the unpaired electron pair $2S^2$ of nitrogen atom generalized by nitrogen and carbon atoms. As a result, the carbon electron is pushed out from the double π-bond C=O and is localized on the oxygen atom with partial transformation of the C=O bond into a single one

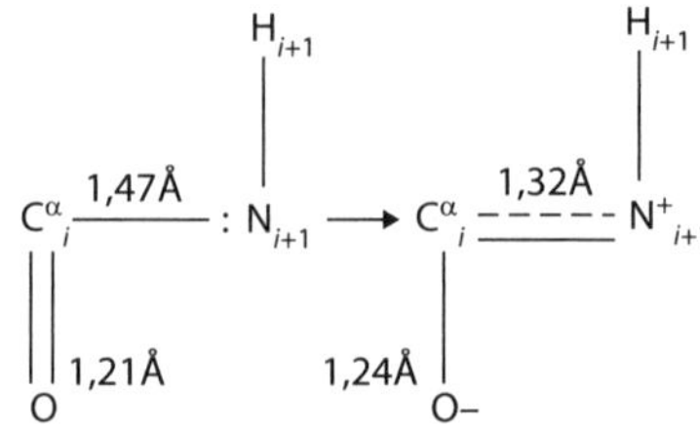

.

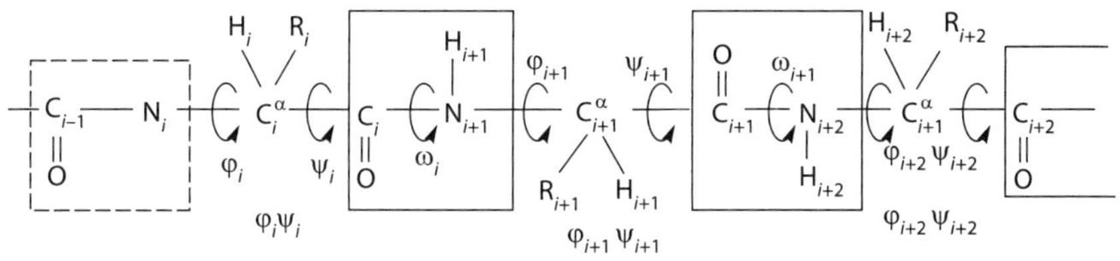

Figure 8.4 Common structure of a polypeptide chain.

Delocalizaion of electrons between N, C and O atoms leads to the maximal stabilization of the peptide group when its atoms, including α-carbon atoms of adjacent amino acids, are positioned in one plane. That is why rotation around peptide bond $C^{\alpha}_{i}{=}N_{i+1}$ is hindered due to its dual nature. Here it is possible to take into consideration only rotation around bonds $N_i - C^{\alpha}_{i}$ (angle φ_I) and $C^{\alpha}_{i} - C_i$ (angle ψ_I) because such a chain has no steric overlapping of atoms of the *i*-th peptide unit with the (*i*+2)-th or (*i*-2)-th units. In other words, only pairwise cooperative interaction takes place in the polypeptide chain upon rotation around single bonds belonging to the same α-carbon atom. Every pair of angles (φ_i and ψ_i) can be analyzed independently, thus cooperativity of the whole chain is in fact determined by interactions of adjacent peptide units. Potentials of internal rotation U_0 around single bonds are quite small (~1.0 kcal/mol). Hence, minima of individual discrete states formed upon changing angles φ and ψ are divided by low barriers. The generalized expression for conformational energy looks like this

$$U(\varphi,\psi) = \sum_{i,k} U_{i,k}(\varphi\,,\psi) + \frac{U^0_{\psi}}{2}(1-\cos 3\varphi) + \frac{U^0_{\psi}}{2}(1-\cos 3\psi) + U_{\text{el.stat}} \tag{8.6}$$

where $U_{i,k}(\varphi,\psi)$ is determined by potential (8.3) with distance $r_{i,k}$ depending on angles φ and ψ. The $U_{el.stat}$ value is estimated by formula (8.4). The conformational energy was calculated (equation 8.6) for the simplest residues of methyl amides of N-acetyl-α-amino acids (CH_3–CONH–CHR–CONH–CH_3), where *R* is the radical of the side chain of the amino acid residue.

Figure 8.5 shows the diagram of the conformational energy surface as a function of angles φ and ψ of the molecule of N–acetyl–L–alanine. One can see four regions of low energy with shallow minima of 1-2 kcal/mol. These estimates of the structure are compatible with other experimental results.

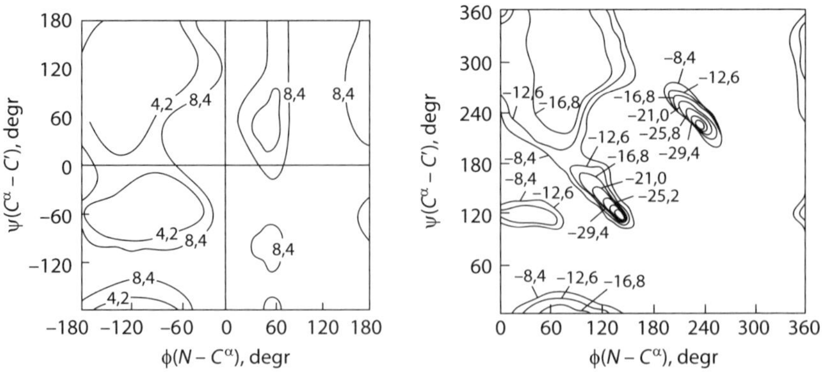

Figure 8.5 Diagram of a potential surface of the molecule of methyl amide–N–acetyl–L–alanine. Numbers indicate conformational energy in kJ/mol (по Попову Е.М., 1981).

Thus, the conformation of simplest fragments of a polypeptide chain can be found from computations. It is necessary to know the chemical sequence of amino acid residues and estimate the energy of non-valence interactions of their atoms and atomic groups using formula (8.6). The obtained conformation is assigned as certain rotation angles of atomic groups and corresponding distances between them, when the total conformational energy that depends on all types of volumetric interactions reaches its minimal values.

But the **conformational energy of a protein** that includes many hundreds of residues cannot be found in this way because of the great mathematical complexity. In this case it is also impossible to calculate directly the secondary and especially the tertiary structures of large regions of protein from its primary sequence. To determine the protein structure, the empirical method is used which is based on voluminous experimental data on the correlation of the secondary structure of a protein region and its primary amino acid sequence. On the ground of these correlations, the heuristic principles of protein three-dimensional structure and the rules of polypeptide chain folding resulted in the formation of the secondary and tertiary structures have been formulated.

The Method of Physical Modeling of protein structure is based on the step-by-step estimation of interactions of atoms without valence bonds between themselves and with the solvent. It is postulated that the protein native conformation consistent with the free energy minimum is characterized by coordination of all types of intramolecular interactions. The latter prevail over intermolecular interactions.

However, direct prediction of the most stable conformation of protein molecule by minimizing the total energy of all types of interactions is impossible because of insurmountable mathematical complexity due to an astronomical number of probable protein conformations and, as a result, of the same huge number of local minima. This makes the search for the general minimum of free energy for the whole protein molecule practically unrealizable.

Thus, if accepted that on average every amino acid residue has ten low-energy forms, the total number of structural elements will be 10^n (where n is the number of residues in the protein chain). A complete search for all probable conformations, even using computer, is possible only for the ~10-membered peptide fragment. Apparently, in this case the only correct way is step-by-step consideration of separate types of interactions, specifying the conformation of every residue with the nearest, neighboring and distant regions in the chain. As a whole, such an algorithm reminds the "block" folding model when separate structural elements of protein, having been formed at the earliest folding stages, do not disintegrate any longer, i.e. the so-called folding nuclei are formed.

Short-range interactions predetermine the set of low-energy conformations for a free monopeptide, while the middle- and long-range interactions determine the choice of such conformations from this set which are involved in the typical native structure of protein. On account of the coherent interactions in protein, the middle- and long-range interactions do not "spoil" the initial advantageous low-energy conformation formed at early folding steps. That is why the low-energy states of free residues should unavoidably be involved in the low-energy states of the fragments. In turn, this suggests the existence of structural peculiarities in free fragments satisfying the middle- and long-range interactions in the entire globule.

In other words, the deterministic mechanism and the cooperative character of self-organization of the protein globule should lead to energy differentiation and emergence of low-energy conformational states at early stages of self-assembly. Hence it follows that at the first stage of calculations it is required to take into account the total set of low-energy states of the free residues, which are later selected to be included in the native structure.

Specific calculations are based on the analysis of atom-to-atom interactions including non-valence, electrostatic and torsion interactions and hydrogen bonds. As an example let us analyze the data on studying the spatial structure and conformational possibilities of fragment *arg*-1-*cys*-38 from the molecule of bovine pancreas trypsin inhibitor (BPTI) consisting of 58 residues (Popov, 1981). At the first stage, the structures of several tri- and tetrapeptides were calculated, which allowed obtaining low-energy conformations of larger fragments (see Fig. 8.6). Further analysis included the overcoming of steric hindrances that appeared upon association of

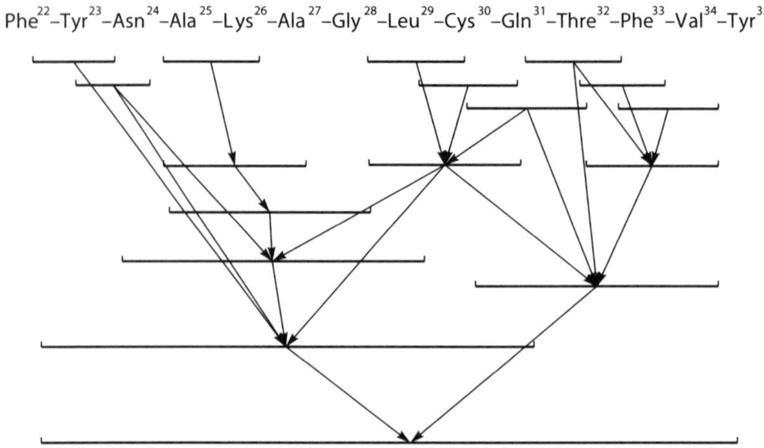

Figure 8.6 Sequence of taking into account interpeptide interactions when calculating the *phe*-22-*tyr*-35 fragment of BPTI (по Попову Е.М., 1981).

different fragments, and association of fragment *phe*-22-*tyr*-35 shown in the figure, with fragments *arg*-1-*tyr*-21 and *tyr*-35-*gly*-37.

The obtained structure was compatible with the experimentally determined geometry of this region with an accuracy of 0.1 nm. Thus such a semiempirical conformational analysis can be successfully used for determining the spatial structure of sufficiently complicated peptides consists of at least up to forty amino acid residues, retaining the initially selected most favorable conformational states of free peptides.

The typical topography of a protein globule is determined mostly by polar groups positions located on the surface, while non-polar groups are mainly located within the globule forming its hydrophobic core. The surfaces of basic secondary structure elements (α-helices and β-structures) have complete hydrophobic regions. The intramolecular hydrogen bonds between peptide groups are maximally saturated and stabilize the globule.

In addition to these topological principles, there are vast statistical data on the frequencies of each amino acid residues from the primary sequence in α- and β-elements of the secondary structure. Using the empirical rules, it is possible nearly in half of the cases to predict what the protein secondary structure would be at the assigned primary sequence.

The naturally occurring types of spatial packing of protein molecules are limited, various groups of evolutionary related proteins corresponding to each of them. In accordance with the available structural classification the identified proteins are subdivided into ≈3500 structural families forming

$\approx$1000 types of spatial packing. Thereby given the protein known to belong to the family of integral receptors of the cell membrane, its structure should contain hydrophobic α-helices piercing the membrane approximately perpendicularly to its plane, whereas polar and charged residues should be concentrated in the "loop" regions (connecting transmembranes) and at the N- and C-termini.

The "affinity" between proteins (typically measured by the degree of identity of their amino acid sequences) is not accidental: one of the accepted hypotheses of protein evolution explains the "relationship" by gene duplication that occurred during organism evolution and resulted in the formation of a protein with a novel function. And although the "novel" protein acquires another function and its sequence gradually evolves and alters, its three-dimensional structure remains rather conservative up to some moment (Lesk and Chothia, 1986).

It was found empirically that if the sequences of two proteins are identical by more than 30%, the proteins are related with a high degree of probability and the extent of evolutionary divergence is not so great yet for their structures to lose common features. Namely these observations make the basis of the technique for predicting the three-dimensional structure called the homology modeling (or comparative modeling).

At the present moment, the homology modeling allows determining the structure of more than half of proteins the structure of which has not been identified so far. The process of homology modeling includes several steps, the central steps being the search for a pattern (a related protein with identified structure) and the amino acid alignment. It has been estimated that following optimal strategy to identify the "target" proteins when at least one structural pattern will be known for each protein, it is sufficient to construct altogether about 16,000 structures in order to identify the structure of the major part (>90%) of the remaining proteins using the modeling by homology.

Alignment of amino acid sequences of two and more proteins is an indispensable tool of bioinformatics. The alignment looks like a set of sequences written one above the other in such a way that it is possible to reveal regions of maximal conformity (homology) of the proteins aligned.

Various types of protein structures form structural hierarchy (Fig. 8.7) which apparently reflects also the sequence of stages of protein folding from the primary polypeptide chain. Already at the very earliest stages of folding in the unfolded chain, α- or β-regions of the secondary structure are formed as a result of local interactions. Then these regions are stabilized due to the action of hydrophobic forces, hydrogen bonds and volumetric interactions with other chain regions now forming the tertiary structure. Self-assembly of the protein structure has a directed cooperative character.

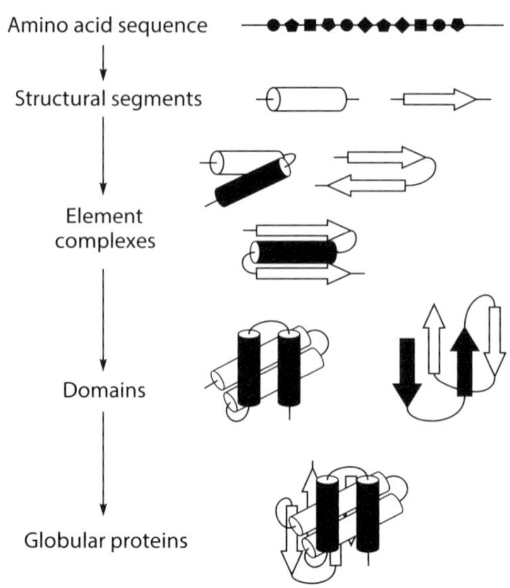

Figure 8.7 Hierarchy of protein structure and sequence of protein folding (Финкельштейн А.В., Птицын О.Б. «Физика белка», 2012)

Protein Folding. As early as in 1973, Anfinsen demonstrated that the protein spatial structure is capable of spontaneous self-organization. After returning to normal temperature conditions and removal of the denaturant from the solution, in several minutes the denatured protein globule restores its initial native structure in which each atom occupies its specific position.

With a great number of conformational substates in native protein ($\sim 10^{100}$ for a chain of 100 residues if each of them has ~10 conformations), the real process of folding should have a directed nature. It cannot be realized by consecutive search for all versions to determine the energy minimum required for the structure stabilization.

Indeed, upon protein chain folding the entropy decreases due to the growth of ordering and concurrently the energy drops as a result of the formation of contacts between the approaching links. In the case of the stable folded structure, the loss due to the decreased entropy ($T \Delta S < 0$) should be compensated by the gain ($\Delta U < 0$) in energy, so that in total $\Delta F = \Delta U - T\Delta S < 0$ ($|\Delta U| > |T\Delta S|$). Assume that the search for this energy minimum proceeds by selection of different types of packing. Then in the process of folding the chain should first approach at random the native conformation having lost almost all entropy, and only after that the gain of energy will take place as a result of the formation of corresponding contacts.

In this case, the process of folding will be very slow because at the first stage the loss of entropy ($-T\ \Delta\ S$) goes along with the corresponding growth of free energy $\Delta\ G^*$ of the transition state formed during folding. In its turn, the duration of the process $\tau \sim \exp\ (\Delta\ G^*/RT)$ depends exponentially on the $\Delta\ G^*$ value too. Namely this circumstance (the entropy decrease prior to the beginning of the energy gain) underlies the "Levinthal paradox" in accord to which the protein chain cannot find the most stable structure for a reasonable time. Let the time of transition of one link from the coil to the growing globule is $\tau \sim 1$ ns, and the entropy decrease by one amino acid residue having 10 conformations is about R ln 10. Then the folding would take on the whole 10^N ns. By Levinthal's estimate, the time of stable structure formation for protein with N=58 residues would make $\sim 10^N$ n $\sim 10^{58}$ ns $\sim 10^{41}$ years.

In the meantime, the time of 10^{-2}-10^2 s is sufficient for natural amino acid sequences to fold spontaneously into a native globule, the stability of which only slightly (~0.1 kcal/mol per chain link) exceeds that of the unfolded form (a rather shallow energy minimum).

From the point of view of thermodynamics, folding requires compensation for the energy decrease at each folding stage (Finkelshtein, 1996).

The sequential process of protein structure folding begins with the fixation of one link just in its final conformation, i.e. with the formation of the secondary structure nucleus. Then the links adjacent in the chain are also fixed in the final conformation, in which structure-stabilizing interactions arise.

In this case the insignificant value $\Delta\ G^*$ of the energy barrier is conditioned by the fact that the entropy decrease upon fixation of one link is compensated by the energy gain at the formation of the "required" contact which is not disrupted later. This drastically diminishes the general time of sequential folding of the chain. Such a mode of protein folding is ensured eventually by that the globular portion of the growing nuclei of protein structure is not rearranged during folding. The assumption complies with the principle of coordination of short- and long-range interactions in protein that underlies its cooperative properties.

There is a standpoint that natural proteins and appropriate amino acid sequences have special properties providing the unique character of their folding. This means that evolutionarily selected natural proteins, the number of which ($\sim 10^5$) is much lower than their total plausible number, are not a result of the "random choice storage". Rather they do possess some fundamental properties that distinguish them from other unnatural sequences.

This is supported by the fact that namely the main sequences selected by nature fold spontaneously into a stable native structure, doing it promptly.

Computations demonstrated that their energy spectrum has a gap separating the native structure from the nearest unfolded state. That is why the native state is thermally stable.

In the cell, the formation of the protein native structure occurs much more quickly than protein renaturation in solution. It does not begin from the state of a stochastic coil as upon renaturation in solution, but yet on the ribosome without the whole chain release into the ribosome environment, i.e. cotranslationally (Spirin, 1991).

Special mechanisms (cotranslational folding, chaperons) fully provide for the realization in the cell of the above physical principles of directed cooperative folding of the protein sequence into a stable native conformation.

Hydrophobic Effect. From the viewpoint of macromolecule structure stabilization, the most intriguing are interactions of water with hydrophobic compounds. Studies of solubility of nonpolar hydrocarbons upon their transfer from nonpolar solvents to water demonstrated that dissolving of these compounds in water is an exothermal process. It proceeds with a temperature rise and thermal energy release, which testifies to the decrease of enthalpy in the system upon mixing hydrocarbons and water ($\Delta H < 0$). But this enthalpy decrease is compensated by the entropy drop ($\Delta S < 0$) and, as a result of this, an increase in the overall thermodynamic potential ($\Delta G > 0$). All together, the absolute value of the enthalpy decrease ($\Delta H < 0$) is lower than the entropy drop, and consequently the dissolving of nonpolar compounds in water is thermodynamically unfavorable: $\Delta G = \Delta H - T\,\Delta S > 0$.

Thus, upon transfer from benzene and dissolving ethane in water at 298 K, the entropy drops by $\Delta S = -84$ J/(mol·K), and the enthalpy by $\Delta H = -9240$ J/mol. The total increase of the thermodynamic potential of the ethane–water system as compared to that of the ethane–benzene system makes

$$\Delta G = \Delta H - T\Delta S = -9240 + 298 \cdot 84 \simeq 16000 \text{ J/mol} = 16 \text{ kJ/mol}.$$

The result of this is the repulsion by water of molecules of nonpolar substances so that it is more favorable thermodynamically to interact with each other than with water.

The detailed molecular pattern reflecting changes of water structure and mutual orientation of molecules of dissolved substance and water as a solvent is quite complicated.

When nonelectrolyte molecules are immersed in water, water molecules should orient themselves on the nonelectrolyte surface to form a maximally possible number of hydrogen bonds in conditions of disrupted regular water cellular structure and disappearance of voids in the boundary region. This causes compression of water molecules, adjacent to the hydrophobic surface, and increases the mean time of the "settled" life to $\tau \sim 10^{-7}$ s.

The formation of a compressed slow-moving layer of water molecules around nonpolar hydrocarbon radicals reduces the entropy of the system. This reduction is not compensated by weak interactions of hydrocarbons and water molecules. Accordingly, if the nonelectrolyte concentration increases, the molecules of the substance dissolved in water associate with each other. It is accompanied by the concurrent release of a portion of water located around the dissolved molecules with a resultant increase in the entropy of the system. Hence the system is divided in two phases: the solvent and the hydrophobic compound. Therein lies the entropic nature of hydrophobic interactions as a special type of binding of nonpolar particles with each other to reduce their overall contacts with water. The forces enabling these interactions have an entirely entropic character. They are long-range as they arise between hydrophobic groups at a distance exceeding the sum of their van-der-Waals radii. The energy of hydrophobic interactions varies between 2 and 40 kcal/mol and grows with an increase in the contact area of nonpolar particles.

The hydrophobic effect plays a significant role in formation of biostructures, being in itself a basic factor of their stabilization. Actually the effect of interactions of protein polar groups with water polar molecules is associated with the predominance of the polar amino acid residues on the surface of a protein globule. But along with this there may also exist interaction through hydrogen bonds of polar peptide bonds (NH...OC) belonging to various regions of the chain in the globule. Individual amino acid residues differ in their hydrophobic properties and can behave as polar or nonpolar compounds.

When estimating the role of various interactions in stabilization of globular proteins, it should be accepted that the character of the native conformation is determined not by some single effect but is a total result of a cooperative finely interbalanced action of a number of energy and entropy factors. The hydrogen bonds formed between polar groups and water and inside the globule are the cardinal factor in providing the stability of separate regions of a protein molecule. They constrain local conformational changes within protein determining the rigidity of the construction and the general character of potential barriers for internal motions of parts of the native structure.

At the same time, hydrophobic interactions between side groups located in separate regions of the main chain play a decisive role in folding of the globule from the main amino acid sequence and in determining its general shape. In both cases water as a solvent plays a great role, facilitating polar interactions due to the formation of hydrogen bonds both on the surface and within the protein macromolecule.

Heterogeneity of the macromolecular structure affects the distribution of water molecules. Apparently, water can exist as a bulk phase of free solvent and bound water its state being dependent on the nature and sites of localization of protein groups it interacts with.

In many respects the physicochemical nature of bound water is specified by its interactions with weakly mobile structural elements. This explains the existence of biopolymers in solutions of several fractions of bound water. Their life-time values lie between those typical of free (10^{-11} s) and strongly bound water (10^{-3}–10^{-5} s) (Aksenov, 1990).

Functioning of proteins is closely related not only to the character of their conformation, but what is more important, to their conformational mobility depending on the presence of water. So, at low-level hydration of α-chymotrypsin preparations, the arising additional contacts between surface dehydrated polar groups enhance the rigidity of the α-chymotrypsin globule and make it lose the enzyme activity in dimethyl sulfoxide. In the very dry samples, up to some critical hydration value, no activity is observed at all. Upon an increase in hydration of the sample, recovery of its activity is an abrupt process within a narrow range of the increased number of water molecules from 170 to 180 per one protein molecule. It is evident that in this range certain degrees of freedom necessary for the enzyme action are activated. Of significance is the fact that the amount of water involved in this process may be less than could be required to complete the formation of the hydration shell (Khurgin, 1980).

In general the protein–water system may be considered as a single cooperative system where changes in the state of both the solvent and the protein fractions are interconnected.

Uniqueness of Protein Structure. Protein has a specific tightly packed structure, formed of the main amino acid sequence as a result of coordinated character of short- and long-range interactions. In this respect protein structure is unique. The question arises: To what extent does the main amino acid sequence determine unambiguously the topology and functional properties of a protein globule?

As known, proteins performing the same functions in different organisms vary in their primary sequence (for example, cytochromes). But their tertiary structures are similar. However, there are proteins which have similar tertiary structures but accomplish different functions. It was also disclosed that the ability to form α- and β-regions with continuous hydrophobic surfaces is inherent not only to natural amino acid sequences of polar and nonpolar groups, but to random ones too. Thereby, to obtain a tightly packed protein globule, it is not required to predetermine absolutely unambiguously its primary sequence, though actual protein functions are specified by a relatively low number of active groups in the active center. As a consequence in the latter case the situation is quite different. In the protein (enzyme) active center containing typically 5–6 residues, it is impossible to substitute any of them without damaging the functional properties. Thus, the combination of indispensable residues in the protein active center should be reproduced absolutely unambiguously retaining only general topological features of the whole globular spatial structure.

Then using formulas of the information theory (6.2 and 6.3) we can estimate the probability of random protein synthesis performed in a single step. First let us determine the amount of information contained in an active center with 5 or 6 amino acid residues (n = 5–6). The total number of possible ways W to construct such an active center by choosing just the required 5–6 amino acids from the total number of 20, is apparently $W = 20^{N} = 20^{5}–20^{6}$. In this center the amount of information is $I = \log_2 W$:

$$I = \log_2 20^{N} = 4.3\ N = 20–30 \text{ bits}.$$

Obviously this is the most minimal amount of information required for the protein synthesis. In addition, in a protein there are actually other indispensable specific regions, containing as a rule no less than N=20–25 amino acids. Then the amount of biologically unique information in a protein increases. It will make the following value

$$I_{\text{B}} = \log_2 20^{N} \approx 4.3 \cdot N \approx 90–100 \text{ bits}.$$

The probability of an accidental single-step synthesis of such a structure is very low:

$$\text{P} = \frac{1}{W} = 2^{-\text{I}_{\text{Б}}} = 2^{-(90–100)} \sim 10^{-30}.$$

This means that the development of a capacious informational protein system could not occur in a single step, but was realized gradually with mandatory fixation of formed structure elements in reproduction processes.

Peculiarities of Three-dimensional Structure of Nucleic Acids. In contrast to proteins, the DNA structure is more stable. Thermal fluctuations do not result in disruption of hydrogen bonds and do not change the interplanar spacings between the bases. In models the rigidity is the cardinal parameter. The double DNA helix has absolute rigidity along the helix length and at the same time a limited number of rotational degrees of freedom around single chemical bonds. DNA conformations belong either to the *A*- or *B*-forms. In the case of *B*-forms, the helix axis passes through base pairs near their centers of gravity, while in the *A*-forms the aperture of about 4 Å remains in the center and the bases are forced out to the molecule periphery. The main difficulty in a comprehensive description of energetically possible conformations of double helices is an extremely great set of all structure variants.

The conformation of a nucleic acid monomer is determined by the conformation of the sugar ring, five dihedral angles of rotation around single bonds in the sugar-phosphate chain and one angle x controlling the orientation of the base relative to the sugar ring. The difference in the *A*- and *B*-forms is that in the *A*-form, the values of both angle x and angle of rotation τ between adjacent pairs vary and the values of distance of the *D* pair from the helix axis are also large (Fig. 8.8). Moreover, the alternative geometry of the sugar ring in the *A*- and *B*-forms is defined by which of the carbon atoms slides out of the plane of the sugar ring. The presence of such a great number of internal degrees of freedom allows considering the process of varying the double helix conformation to be permanent. But it is also possible not to go beyond merely "reasonable" mutual positions of the base pairs. In this case it turns out that only a limited number of

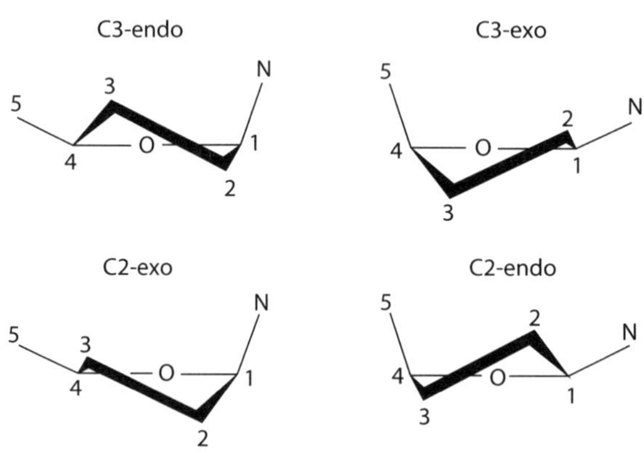

Figure 8.8 Forms of DNA helix: four most stable conformations of the nucleotide sugar ring.

conformations of the sugar-phosphate backbone exist for a regular helix. Estimations show that the energy of *A*-forms is on the whole larger and the width of the energy well is narrower than those of *B*-forms. This is the result of the fact that the space between similarly charged phosphates of the same chain is by nearly 1 Å shorter in *A*-forms. Thorough consideration of all interactions shows that the *B*-form is the only stable form of DNA.

Lately data have been obtained on the existence of the so-called *Z*-form of DNA in which the helix with antiparallel filaments is twisted left-handedly, and the iteration unit has two nucleotides rather than one.

9

Intramolecular Dynamics of Proteins

The fact that functional properties of proteins depend on their conformational state has been long recognized in biochemistry. But it has remained unclear what forces and mechanisms govern changes of this state and what is the role the latter plays in protein activity. In modern molecular biophysics, the fundamental problem of the mechanism and functional role of intramolecular mobility of biopolymers and in the first place proteins holds the central position.

To solve this problem it is necessary to obtain qualitative characteristics of the mobility of separate protein parts and atomic groups related to the functional state of a protein molecule under different conditions (temperature, pH, and ionic strength). Due to protein heterogeneous three-dimensional organization, movements of its individual structural units would differ in their characteristic times and displacement amplitudes. Moreover, deep conformational rearrangements of a whole protein molecule cannot occur in a single act but rather include local microinformation displacements of separate atomic groups. As a result, they lead to directed rearrangements of the whole structure of a protein molecule via mutual displacements of its separate parts in line with specific deterministic degrees of freedom. Let us consider conformational changes of hemoglobin (Hb) as an example of sequential, functionally significant rearrangements in a protein molecule. Oxygen binding to an iron atom Fe^{+2} triggers its displacement by 0.7 Å into the plane of the heme group. This primary microinformation displacement induces a cascade of consecutive rearrangements in hemoglobin. They include the shift of histidine to the center of the molecule and displacement of subunits accompanied by the rearrangement of the system of hydrogen bonds. At the same time the conformation of α- and β-subunits themselves is changed as well. After oxygenation of the initial two α-subunits the bending of the subsequent oxygen molecules with the remaining subunits is facilitated since their affinity to oxygen enhances several hundred times. Actually this circumstance determines the functional meaning of cooperative conformational rearrangements proceeding in hemoglobin. The physical reason of such rearrangements is that the initial balance of forces becomes irreversibly distorted as a result of oxygen binding to Fe^{+2}. Thus a new equilibrium oxyconformation, matched by another spatial arrangement of atomic groups, should be formed in a hemoglobin molecule. To reach this final equilibrium oxyconformation a system passes over a number of consecutive stages starting from the relaxation of the original desoxy form that has become non-equilibrium after fast association of oxygen with Fe^{+2}. Direct investigation of fast internal motions within a protein molecule has become possible only recently thanks to the state-of-the-art physical resonance methods of radiospectroscopy (electron paramagnetic, nuclear magnetic, and nuclear gamma-resonances and luminescence). Let us briefly analyze the physical nature of these methods and information obtained using them.

The basic experimental approach is to determine certain physical parameters (luminescent and paramagnetic) of low-molecular compounds incorporated into a protein molecule and thereby characterize the mobility of protein surrounding in the vicinity of these compounds , i.e. to obtain the characteristics of protein intramolecular mobility.

Luminescence Methods permit measuring intramolecular mobility of protein by studying temperature dependence of the shift of the luminescence maximum of the label, incorporated into a protein, or the shift of the selfluminescence maximum of tryptophan residues in a protein. When a luminescent molecule absorbs a light quantum, one of the two π-electrons transfers to the excited singlet level S_1 so that the whole molecule passes into the excited singlet state. Upon transition to the excited state the store of vibrational energy in the molecule momentarily increases and then in 10^{-11}–10^{-12} s vibrational energy is dissipated with the relaxation of the molecule to the lower vibrational sublevels in state $S_{1.}$ The excited state S_1 of the

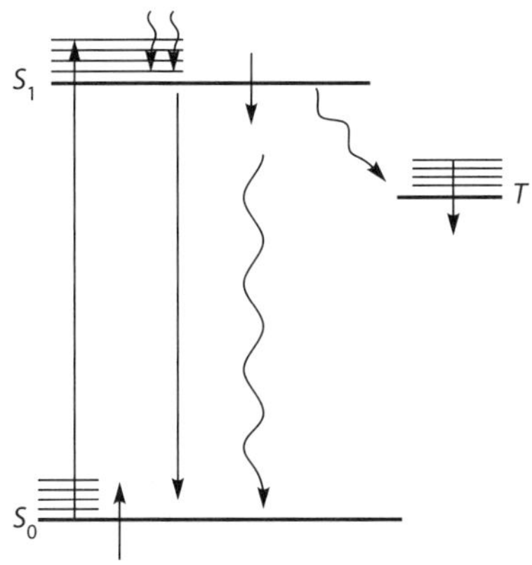

Figure 9.1 Scheme of transitions between singlet (S_1 and S_0) and triplet (T) states of a molecule.

molecule exists for $\tau^* = 10^{-8}–10^{9}$ s. After that the molecule returns to the ground level S_0 either with the emission of a fluorescence quantum or without light emission by dissipating electron excitation into heat. During the period τ^* of the state S_1 existence the electron spin on level S_1 can change its orientation to the opposite.

Then it would be parallel to the spin of the other π-electron it was originally coupled to. In this case the molecule performs the transition into triplet state $S_0 \rightarrow T$ where electron spins on S_0– and T–levels are parallel (Fig. 9.1). Now the transition to the ground state $T \rightarrow S_0$ again demands the antiparallel reorientation of spins. That is why the probability of the $T \rightarrow S_0$ transition is low, and the lifetime of state T is great as compared to state S_1 and makes $10^{-6}–10^{-2}$ s and higher. The molecule moves from the triplet level to the ground one with the emission of phosphorescent light. While the molecule remains in excited state S_1 or T, its electric dipole moment affects the dipoles of the environment. Thus during the lifetime of the excited state the dipoles of the environment may have time to reorient themselves in compliance with the excited dipole field. This will happen if the time of the dipole relaxation of the environment τ_p is much lower than lifetime τ^* of the excited state ($\tau_p << \tau^*$). And accordingly the energy level of the excited molecule will decrease so that the maximum of its fluorescence spectrum shifts to the long-wave length region. But if $\tau_p >> \tau^*$ there is not enough time for the reorientation of dipoles of the environment, i.e. in this case the surrounding of the fluorescent molecule is rigid and does not affect the position of its fluorescence maximum. Valuable information on intramolecular mobility of proteins is obtained by analyzing in this way tryptophan self-fluorescence. The position of the fluorescence maximum strongly depends on the mobility of the dipoles of the tryptophan in a protein environment and may vary up to 30 nm. In the complicated protein system the time of reorganization of the tryptophan environment can exceed the time of its electron S_1 excitation state ($\tau^* \sim 5$ ns) in particular at low temperatures. In fact it has been found that a temperature decrease from 0° to −20° entails a shift of the tryptophan fluorescence maximum by 5–12 nm while a temperature decrease from −20° to −90° causes the shift of by 4–9 nm to the short-wave length region of the spectrum. This testifies to the freezing of motions in the protein matrix in the nanosecond range. In lyophilized proteins, no such shifts are observed which evidences that the presence of water in protein structure is necessary to provide for its mobility. The protein–water system is frozen as a single microphase. If exogenous phosphorescence labels (for example, eosin derivatives) are incorporated into protein, this would allow estimating longer characteristic times of structural rearrangements (up to 1 s). It was found that in deep layers of protein macromolecules, slow motions ($\tau_p \sim 1$ s) take place, while in layers closer to the surface such motions are much faster ($\tau_p \sim 10^{-3}$ s). One should clearly apprehend that the determined times characterize the mobility of only the nearest microenvironment of a luminescent chromophore and cannot be applied to the whole macromolecule. The number and sites of localization of such luminescent labels in protein are determined by chemical peculiarities and the character of interactions with surrounding protein groups.

Radiospectroscopy Methods include mainly the method of electron paramagnetic resonance (EPR) and nuclear magnetic resonance (NMR). The physical bases of these methods are quite intricate and are described in detail in special manuals and textbooks. Herein we will only briefly discuss the problems required to understand their application in biophysics. Both electrons and atomic nuclei have their own magnetic moment or spin. If a system of such spins is located in a constant external magnetic field $\vec{H}_0$, the spins will be oriented along the field vector. The energy level will split in two levels. In this case, the energy of the electron with spin $S = \frac{1}{2}$ will depend on whether the spins are oriented parallel (high level) or antiparallel (lower level) to the direction of field $\vec{H}_0$ (Fig. 9.2). Accordingly, the lower level will be occupied by proton nuclei (spin 1/2 with parallel orientation to the field), and the high level will be occupied by spins with antiparallel orientation to field H_0 (Fig. 9.2).

The difference in the energy levels is

$$E_2 - E_1 = \Delta E = g\beta H_0. \tag{9.1}$$

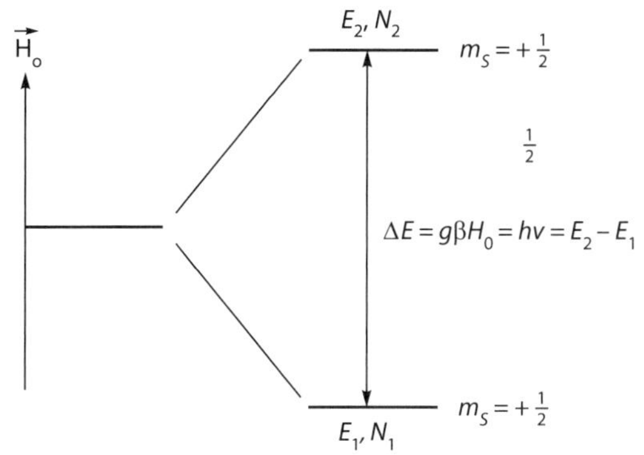

Figure 9.2 Splitting of energy levels of electron (proton) in the magnetic field (H).

In the above formula, g is a constant value, the so-called g-factor which for a free electron makes 2.0023 and for a proton nucleus 5.58. The β value is the Bohr magneton, which for an electron makes ~10–21 erg/gauss, or nuclear magneton $\beta_N = 5 \cdot 10^{-24}$ egr/gauss. At the lower level E_1 the population (N_1) of particles is higher than on the high level (N_2), the proportion of population numbers depending on the difference of energies $\Delta E = E_2 - E_1$:

$$\frac{N_1}{N_2} = e^{g\beta H_0 / K_B T} = e^{\Delta E / K_B T}. \tag{9.2}$$

Let us in addition to constant field $\vec{H}_0$ also apply alternative magnetic field with perpendicular to the constant field frequency (ν). Then induced transitions between the two levels $N_1 \rightarrow N_2$ and $N_2 \rightarrow N_1$ will occur in the system. But since initially $N_1 > N_2$, the number of transitions from the lower level to the high one $N_1 \rightarrow N_2$ will exceed the number of backward transitions $N_2 \rightarrow N_1$. As a result we observe resonance absorption of energy of the alternative field if

$$h\nu = g\beta H_0. \tag{9.3}$$

When the external field has the value of $\vec{H}_0 = 10^4$ gauss, the resonance frequency for a free electron is $\nu_e = 2.8 \cdot 10^{10}$ s^{-1}, while for a proton nucleus it is much lower $\nu_H = 2.26 \cdot 10^7$ s^{-1}. Figure 9.3 shows the absorption line of the alternative field and its first derivative (EPR signal) versus the value of the constant magnetic field H_0. Usually as far as constant frequency ν of the alternative field is set and the existence of the external constant field is

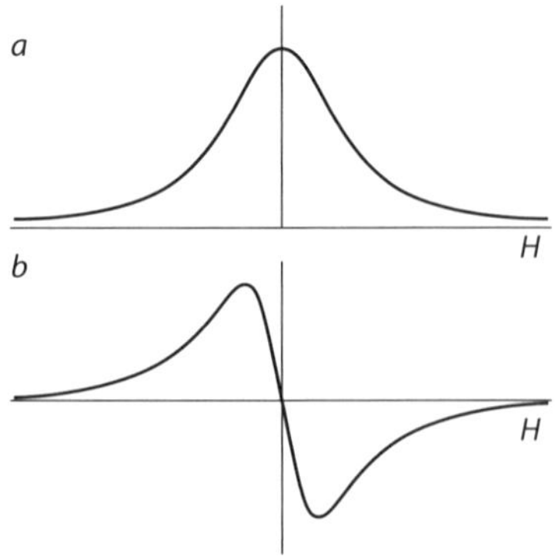

Figure 9.3 Absorption line of microwave frequency of the field (a) and its first derivative (b).

smoothly changed to attain those values of H_0 when resonance absorption takes place (9.3).

As seen in Fig. 9.3, the EPR signal has an absorption band of a definite width rather than a narrow line. This means that in fact the energy levels of the spin state themselves are somewhat blurred, i.e. they have a specific Gaussian width. The major reason of this is that the interaction with external magnetic fields is not the only process of energy exchange in the system of oriented spins. The system is not isolated but it interacts with the environment and transfers its energy at an averaged time T_1. This prevents equalization of populations N_1 and N_2 under the action of the alternative field. There are also other processes of energy exchange in the system of spins *per se*, which influence each other and alter the energy of the spin states. Spin-spin interactions generate additional local magnetic fields in the sites where neighboring spins are localized and thus change the strict resonance condition (9.3) widening the absorption band. Time T_2, when the relative energy of the given spin state is still retained, depends on the spin-spin interactions. Apparently, time T_2 of the retention of the initial spin state is the greater the lower spin-spin interactions. The described relaxation processes of the initial spin state decay result in that the actual position of spin energy level E_0 becomes somewhat uncertain and corresponding width Γ.

In quantum mechanics, the relationship between lifetime τ of the given state and the observed width Γ of its actual energy level position (E_0) is determined by the relation of uncertainty for the energy

$$\Gamma \cdot \tau \sim \hbar, \tag{9.4}$$

where $\hbar$ is the Planck constant h divided by 2π ($\hbar = h/2\pi$). So, the actual position of the energy level (E_0) of the electron state in the system with relaxation time T_1 can be determined from the position of the resonance line with the accuracy up to $\hbar/T_1$ $\left(E_0 \pm \frac{\hbar}{T_1}\right)$. At $\tau \to \infty$ the width $\Gamma \to 0$ and the energy E_0 value is determined precisely. In a general case, the resonance line width is determined as

$$\Delta H \cong \frac{1}{2T_1} + \frac{1}{T_2}. \tag{9.5}$$

For free radicals, $T_1 >> T_2$ and the EPR line width is dependent mainly on time T_2:

$$\Delta H \sim \frac{1}{T_2}. \tag{9.6}$$

Our main interest is how the observed width Γ is related to intramolecular dynamics in a protein molecule. If the internal viscosity of the protein is low, then due to fast molecular motions spin interference has enough time to be averaged upon absorption of the alternative field. In this case, the resonance is observed at an averaged line with a reduced width. On the contrary, freezing of the solution and reduction of its viscosity would again slow down internal molecular motions. Then upon absorption of the energy of the alternative field spins have time to experience the effect of different local fields in their microenvironment, which will reduce time T_2 of the spin state retention and will bring about the widening of the general resonance line. So, by measuring the absorption line width it is possible to get data on the character of intramolecular motions in the microenvironment of a spin particle (radical, proton). When the method of spin labeling is used, a free radical generating an EPR signal is attached to the functional protein group. The characteristics of the EPR signal are indices of the mobility of the microenvironment of this radical. Among the most suitable labels are nitroxyl labels containing free-radical group $N-\dot{O}$ where the uncoupled electron is actually delocalized between N and O atoms. The EPR signal of this label has a specific structure caused by the influence of the magnetic moment of the nitrogen nucleus on the uncoupled electron in addition to the action of the external magnetic field. The rotation of the nitroxyl radical relative to the direction of the external field changes the structure of the EPR signal. There are calculating methods which allow estimating the characteristic time τ_c of the label rotation using the determined parameters of the observed EPR signal. It turned out that the rotation time of the label depends on the depth of its submersion into the layers of a protein globule. In the external water layer, the τ_c values make 10^{-11}–10^{-10} s. Surface layers are characterized by $\tau_c \sim 10^{-10}$–10^{-8} s, and finally in deep layers of the dense hydrophobic core of the globule $\tau_c \sim 10^{-8}$–10^{-7} s. The enzyme activity of lyophilized preparations of reaction centers of photosynthesis is drastically enhanced upon their moistening, which correlates with a concurrent drop in the τ_c values of the spin label. This testifies to the changed dynamic state and simultaneous emergence of intramolecular mobility and functional activity of protein.

NMR Method. Investigations of the dynamics of protein structure with the NMR method are based on measuring the relaxation times T_1 and T_2 by the width of the absorption line. These data allow determining the τ_c value for nuclei that are in resonance.

By studying the proton resonance it is possible theoretically to estimate the mobility of protein groups which include "resonant" protons. The NMR method allows analyzing specific types of intramolecular motion in proteins. Thus, temperature changes in time T_1 in serum albumin reveal rotations of methyl groups with times $\tau_c \sim 10^{-10}$ s, and at $-100°$ slower motions are observed with $\tau_c \sim 10^{-8}$ s (polypeptide chain oscillations may occur). Measuring time T_2 it is possible to find out three characteristic ranges of time τ_c for proton resonance: slow rotation of a globule as a whole ($\tau_c \sim 10^{-5}$ s), internal motions ($\tau_c \sim 10^{-5}$–10^{-6} s), and movements of external most mobile protein groups ($\tau_c \sim 10^{-10}$ s). The NMR method also allows us to obtain important information on chemical structure of a molecule. It is revealed as a result of various effects of local fields of adjacent nuclei on the position of the resonance band of separate protons occupying different sites in the molecule structure. As a consequence, resonance on different protons is observed at diverse frequencies. This event is called a chemical shift. Moreover, structure of an NMR band, as in the case of EPR, reflects the presence of magnetic moments of nuclei, fine effects of their influence being transferred through the electron bonds in the molecular structure.

Figure 9.4 shows the spectrum of proton resonance of acetaldehyde CH_3–CHO. One can see two groups of lines from three protons of the chemical methyl group and from one proton of the aldehyde group in the molecule. The two groups are separated by a chemical shift. Besides, we can see splitting of each group into individual lines because of the influence of magnetic moments of neighboring nuclei on each other. It should be noted that such decoding of the NMR spectra of protein molecules is complicated by their complex character and overlapping of the basic group of resonance lines of protons. At present the data on ^{31}P

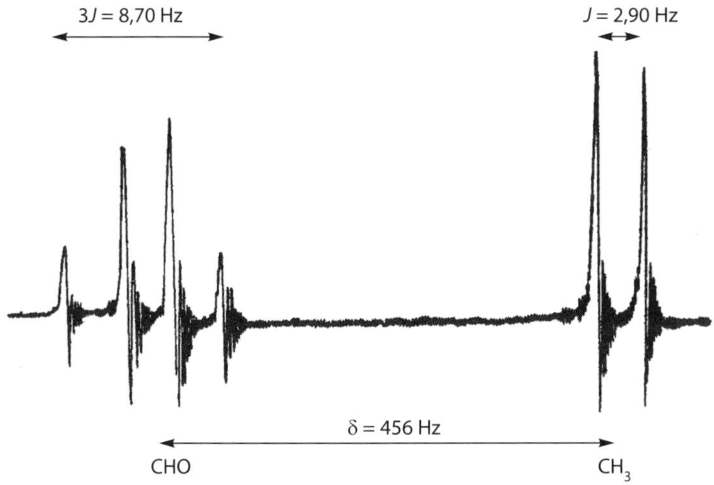

Figure 9.4 NMR spectrum of acetaldehyde CH_3–CHO. The magnetic field grows from the left to the right. *I* is the value of superfine splitting (Carrington A., McLachlan A.D. // Harper & Row, publishers, 1967).

resonance are widely used as well as on ^{13}C resonance since their lines are located at some distance from the main group of overlapping lines in NMR spectra.

Method of Nuclear Gamma-Resonance (NGR) Spectroscopy provides essential data on the protein dynamics. As distinct from EPR and NMR methods, it allows us to obtain not only time characteristics, but also amplitude characteristics of motions in protein. The method allows determining average values of atomic displacements in protein structure that occur during short periods of time 10^{-7}–10^{9} s. The method is based on resonance absorption of γ-quanta by heavy atomic nuclei, for example, the nucleus of isotope ^{57}Fe from natural compounds where its contents make 2.2%. The energy of the γ-quantum absorbed by the ^{57}Fe nucleus makes $\Delta E = 0.0144$ MeV, and the lifetime of the ^{57}Fe nucleus in an excited state is $\tau^* \sim 10^{-7}$ s. Hence, using the uncertainty relation for energy (9.4) it is possible to determine that the natural line width of γ-quanta absorption is very small: $\Gamma \sim 10^{-8}$ eV. However, under actual conditions, the resonance frequency of γ-quanta absorption by the nucleus does not coincide with that of the γ-quantum itself. The point is that upon absorption of a γ-quantum by the ^{57}Fe nucleus, part of the quantum energy is converted into kinetic recoil energy, that is into translational energy of motion of the nucleus itself. This means that actually the energy of the γ-quantum used to excite directly the ^{57}Fe nucleus will decrease by the value equal to the recoil energy. As a result, the resonance absorption frequency will be shifted as compared to the initial frequency of the γ-quantum itself. As the natural width of the line is too narrow ($\Gamma \sim 10^{-8}$ eV) and the recoil energy for free ^{57}Fe is much larger ($2{\cdot}10^{-3}$ eV), this shift will be especially noticeable. The actual value of the recoil energy depends on the capability of the ^{57}Fe atomic nucleus to accept the recoil impulse, which in its turn depends on its own mobility. In solid matter the absorption of γ-quanta may become recoilless, when the entire crystal as a whole takes up the recoil energy so that no displacements of the ^{57}Fe nucleus are observed. Such recoilless absorption of γ-quanta is called the Mossbauer effect. In this case, the spectra of γ-quanta absorption have lines without energy shifts due to the lack of recoil. Upon an increase in the root-mean-square shift $\overline{x}^2$ of the Mossbauer ^{57}Fe nucleus the probability f' of recoilless γ-quantum absorption is reduced. On the contrary, at small $\overline{x}^2$ values no recoil energy is delivered to the nucleus rigidly associated with its surrounding, and thus the probability f' approaches a unity. The dependence of the probability of recoilless absorption on the $\overline{x}^2$ values is exponential:

$$f' = e^{-\overline{x}^2/\lambdabar} \tag{9.7}$$

where $\lambdabar = \lambda/2\pi$ and λ is the wavelength of the γ-quantum (λ=0.13 Å for ^{57}Fe).

By determining the probability of recoilless absorption under different temperature and humidity conditions, it is possible to estimate the displacements of the Mossbauer atomic nucleus, i.e. to obtain the amplitude characteristic of its mobility. A typical protein has a specific temperature dependence $f'(T)$ with a bending point which testifies to the cooperative character of defrosting the internal mobility at the bending temperature point. A dry protein does not demonstrate such properties but rather resembles a standard solid in which the $\overline{x}^2$ values increase monotonically. At temperatures of 77 to 200 K the Mossbauer nuclei oscillate with small amplitudes (0.1 Å), while at high temperatures much greater large-scale shifts appear (0.3–0.6 Å). The enhancement of the protein internal mobility in this case also correlates with the corresponding increase in its functional activity (Fig. 9.5). The time resolution of the NGR method is limited by time τ^* of the excited state of the ^{57}Fe nucleus ($\tau^* \sim 10^{-7}$ s). The NGR method is actually applicable for measuring the displacement of the Mossbauer nucleus that can be attained in 10^{-7} s. It is clear that at other times internal displacements of the amplitude will be different. The general classification of the mobility of protein structure elements by the characteristic times of individual types of motions is given in Fig. 9.6.

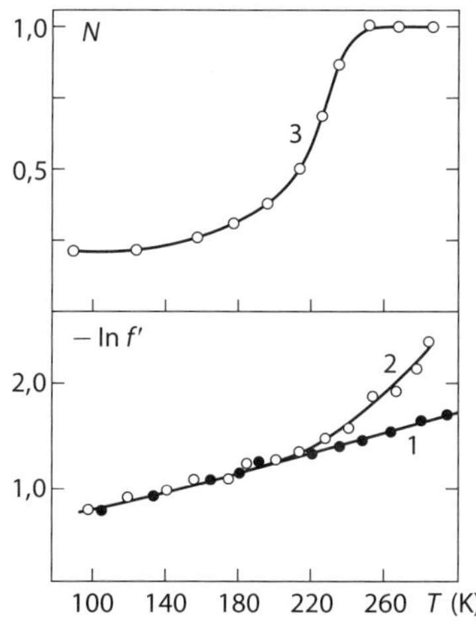

Figure 9.5 Temperature dependence of the probability of the Mossbauer effect ($-\ln f' =< x^2 >$): 1, dry protein; 2, moist protein; 3, functional activity N of the electron transfer in bacterial reaction centers (A.A. Kononenko et al. Chem. Physics, 1986).

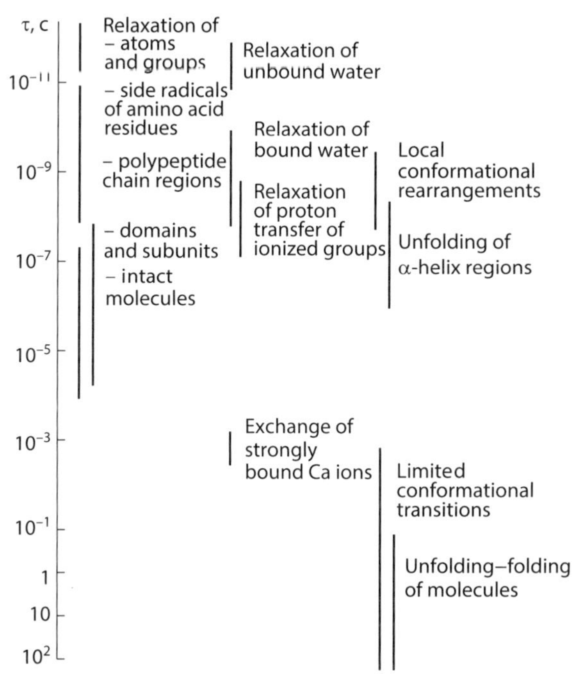

Figure 9.6 Scheme of characteristic times of protein mobility.

10

Physical Models of Protein Dynamic Mobility

In the previous chapter we have described the characteristic times and amplitudes of protein region motions that can be determined experimentally. Let us once again pay attention to the fact that the obtained values depend on the time resolution of the method used and describe the motion in the microenvironment where a corresponding marker is localized. To present a general picture on the basis of these data and understand the internal protein dynamics, it is necessary to construct consistent physical models based on experimental data and analyze their properties. As known, small oscillations of atoms in a solid occur with high frequencies $\omega_0 \sim 10^{13}$ s^1 ($\tau \sim 10^{13}$ s) and low amplitudes $x_a \sim 0.01–0.1$ Å. These oscillations can be described as typical harmonic oscillations. But such an approach is inapplicable for the description of protein micromotions with amplitudes $x_a >> 0.1$ Å. Protein density is very high and comparable with the density of liquids and organic crystals. Therefore a separate protein fragment can be displaced at the distance exceeding 0.1 Å only if the shift is concurrently accompanied by the formation of a fluctuation cavity due to the displacement of other molecular groups surrounding this fragment. It seems as though the protein fragment "pushes away" the neighboring groups. Such motion requires

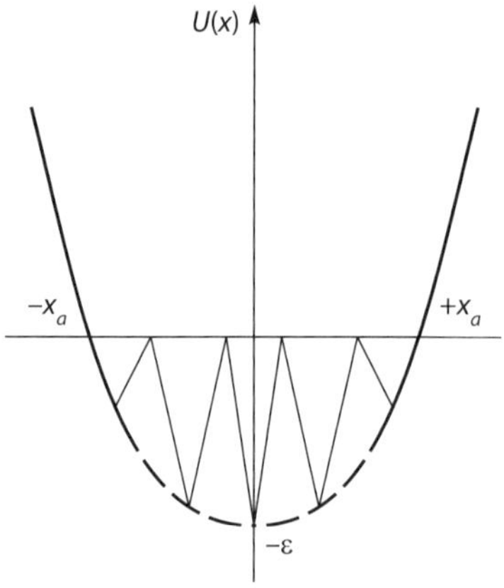

Figure 10.1 Conformational motion in viscous medium (the potential comb imitates viscosity). x_a is the average amplitude of conformational fluctuations; *U(x)* is the conformational potential; *e* is the energy of activation of conformational motions.

the activation energy to overcome potential energy barriers hindering the fragment shifting. The transition from one microstate to another is accompanied by considerable displacements (~ 1 Å) and consequently the barrier width should be large enough. This transition is not similar to a "single-action" jump between two microinformation states. Here the transition through a wide barrier represents a continuous process of motion in the potential field with a complex relief – a thick potential comb (Fig. 10.1).

Thus, the motion of the protein fragment is characterized by two spatial scales. The motion along the "fence" of thick potential barriers corresponds to the interactions of the fragment with the surrounding protein groups. By its nature it is a diffusion process characterized by the conformational diffusion coefficient $D(x)$ which depends on the surrounding and on the conformation coordinate x. The other spatial scale reflects the overall slow displacement along the conformational coordinate that occurs in the conformational potential $U(x)$. As seen in the figure, transition from the conformational energy well $[-x_a, x_a]$ is forbidden because the fragment linked chemically to the remaining protein molecule cannot move away anyhow far arbitrary. Generally such motion complies with the continuous limited diffusion in a viscous medium when the fragment is exposed to random hits or thermal noise. The time dependence of the mean square shift $[x(t)]^2$ in this process is determined from the formula

$$\left[\overline{x}(t)\right]^2 = \overline{x}_a^2(1-e^{-t/\tau c}). \tag{10.1}$$

Here $\overline{x}_a^2$ is the mean square of the shift amplitude and τ_e is the characteristic relaxation time during limited diffusion dependent on the friction upon motion. It is

$$\tau_c = \gamma/m\omega^2{}_0 \tag{10.2}$$

where γ is the friction coefficient proportional to the protein microviscosity, m is the fragment mass, and ω_0 is the oscillation frequency. It should be reminded that the friction coefficient depends on the protein microviscosity in accordance with the Stokes equation

$$\gamma = 6\pi\, b\, \eta, \tag{10.3}$$

where b is the characteristic linear dimension of the fragment (b ~ 1–10 Å) and η is the viscosity in poises. A temperature rise leads to exponential reduction of the viscosity

$$\eta(T) \sim e^{e/k_B T} \tag{10.4}$$

(ε is the activation energy of a viscous flow).

Subsequently a temperature rise reduces exponentially also the time τ_c, because

$$\begin{aligned} \tau_c \frac{\gamma}{m\omega^2{}_0} &= \frac{6\pi\, b\eta}{m\omega^2{}_0}, \\ \tau_2 \sim \eta &\sim e^{e/k_B T}. \end{aligned} \tag{10.5}$$

The enhancement of the medium viscosity results in the growth of τ_c, i.e. in a decrease of the diffusion rate. Hence it is possible to appreciate the temperature influence on the intramolecular motion in protein and, in particular, on the dependence of the mean-square shift of the Mossbauer atom versus temperature (see Fig. 9.5). At low temperatures, the τ_c value is very high, and upon γ-quantum absorption the Mossbauer nucleus has no time to make the shift while it remains in the excited state. In this case the γ-quantum recoilless γ-quantum absorption takes place ($f' \sim 1$). On the contrary, at high temperatures where η and τ_c are small, the nucleus manages to complete the shift and f' drops. Thereby with a temperature rise at the breaking point on the temperature curve (see Fig. 9.5) the τ_c value changes dropping below the critical value. Using the temperature dependence $f'(T)$ it is possible to determine the microviscosity value η for proteins, activation energies ε of the viscous flow, and amplitude x_a of conformational motions. The processing of experimental data for proteins yields typical values $\eta \sim 10^2$ at 300K, which are much higher than those for the viscosity of water and glycerol (ε – 20 kJ/mol and $x_a \sim 0.4$ Å). The α- and β-elements in the protein structure undergo limited diffusion motions depending on the rigidity and microviscosity of the medium. Bending fluctuations of α-helices have a definite form, the amplitude and relaxation time depending sharply on the linear dimensions of the helix. In real conditions, amplitudes of bending fluctuations can be as high as several angstroms with the relaxation times in the microsecond range.

So, the protein has a number of bound elements with essentially different relaxation times. The most fast and small-scale fluctuations are inherent to side groups. These groups form the liquid-like surrounding around helical regions of the polypeptide backbone and play the role of a damping medium. The hierarchy of the relaxation times allows one to represent the protein globule dynamics as fluctuations in a liquid-like drop reinforced with a rigid polypeptide backbone. The dependence of the reaction rate in proteins on the solvent viscosity can be perceived based on these ideas. Diffusion of ligands within the globule occurs only upon the formation of fluctuation cavities or "holes". The appearance of a "hole" inside the globule can be initiated by its formation at first in the solvent on the globule surface. The probability of this process is inversely proportional to the solvent viscosity. Because of conformational motions the surface protein group fills up the "hole" in the solvent. Thus the "hole" now originates in the external layer of the protein. Then, as a result of motions of groups from the second and third layers, the "hole" diffuses within the globule, generating additional fluctuation cavities. As we have seen, the form of these cavities in the protein is not arbitrary, but has the shape of fluctuating slots the parameters of the latter being determined by the geometry of rigid elements of the protein backbone. Solution of diffusion equations permits estimating the velocity of diffusion of particles through such fluctuating slots. Thus, upon diffusion in myoglobin the CO ligand should pass through several "gates" which are open due to conformational motions. The diffusion velocity depends on the ligand diameter, the fluctuation amplitude and the relaxation time of the slot, which in its turn is specified by the rigidity and microviscosity of the walls. For myoglobin, the estimated and experimental data were compatible. In particular, it was found that the total time of CO transport through myoglobin is 10^{-7} s and agrees with the summation of the times of conformational relaxations of several gates in the globule. Note that in a rigid protein molecule where intramolecular motions and structural fluctuations are lacking, the ligand diffusion should be associated with the overcoming of large activation barriers (up to 100 kcal/mol). These barriers would retard the ligand motion in such a way, that it would become infinitely slow at the biological time scale. In other words, in actual biopolymers with dense packing it is structural fluctuations that make possible the transfer of ligands within the molecule, which is important for functional activity.

More complicated types of cooperative relaxation processes in protein will be considered below in connection with the mechanisms of enzyme catalysis.

Numerical Simulation of Protein Dynamics is a modern trend in molecular biophysics. In preceding chapters it has been shown that the protein conformational energy is determined by the atom-atom interactions described by special potential functions. Therefore it is possible to obtain energy maps showing average coordinates of atoms corresponding to the minimum of the total conformational potential. A similar result can be also obtained with the method of X-ray analysis which is used to determine the average statistical positions of atoms in the protein structure. But such methods cannot be used to trace motions and fluctuations of positions of individual atoms underlying conformational fluctuations and transitions in proteins. The method of numerical modeling of protein dynamics for individual atoms allows resolving directly classical equations of motion in which the driving forces are determined from known potential functions of atom-atom interactions. Initial coordinates for heavy atoms (not hydrogen ones) are specified by X-ray data, and at the zero time the rates of their motion are taken to be equal in different randomly chosen directions. Initial accelerations of motion are estimated as first derivatives of the forces acting on atoms in the initial structure. Then, computer-assistant solving of the system of equations for motions of all atoms is performed. The numerical solving of these equations represents trajectories of motion of individual atoms and allows one to obtain a total pattern of structural fluctuations developing with time. It is yet possible to trace atom motions within a short period of time from 0 to $\sim 10^{-7}$ s which is conditioned mainly by time consuming.

Let us take as an example the results of modeling the internal dynamics of trypsin inhibitor (TI) of pancreas, i.e. the protein whose molecule contains 454 heavy atoms. It has turned out that relative to the time-averaged structure, real fluctuations of atom positions in the protein make 0.6 Å for α-carbon atoms and 0.75 Å for all other atoms. Fluctuations of the values of dihedral angles of rotation φ and ψ in the polypeptide chain are also observed varying from 10 to 20° and for angle ω from 7 to 9°. These fluctuations of positions are rapidly damped in 1-2 picoseconds. However there are also long-living (up to 20 picoseconds) fluctuations in positions of α-carbon atoms which evidently reflect conformational transitions in protein. The regularity of fluctuating motions is broken the more noticeably, the more frequently the group collides with other atoms in its microenvironment. Within the limit of total wide conformational minimum the protein experiences spontaneous transitions from one microstate to the other due to thermal energy, such as for example the rotation of the tyrosine aromatic ring in the TI molecule. Computer simulation of this process demonstrated that the transition itself over the potential barrier occurs spontaneously and not because of strong activation collisions with atoms of the microenvironment of the ring. The tyrosine ring overcomes the barrier along a definite trajectory approximately in 1 picosecond, whereas the hits from the microenvironment surrounding are only striving to "draw" the ring aside the barrier and "knock it off" the natural trajectory of the spontaneous transition. Fluctuations in the positions of individual atoms in protein correlate with each other which may lead to larger-scaled structural shifts and conformational rearrangements. The fluctuational trembling of atoms provides conditions and base for functionally directed conformational transitions in proteins. Yet we are still far from constructing a detailed model of the protein dynamics. But even now it is possible to make some generalized conclusions based on a comparison of theoretical and experimental results on intramolecular mobility and its role in the protein functional activity. The protein atomic groups experience the action of various forces (Coulomb and van-der-Waals interactions) as well as random "thermal" hits from adjacent groups. Furthermore, they can be involved also in normal oscillations when small atomic shifts from the equilibrium position have a harmonic character. Comparatively rare "jumps" to overcome activation barriers are also detected. In general every atomic group may be involved in different motions; hence the atom total shift depends in a complicated manner on the contributions of various forces. In most proteins, the fluctuation amplitude grows upon moving away from the molecular center to its periphery. For atoms of the main chain, the mean square shifts make 0.6 Å and for atoms of the long side chains they are about 1.5 Å. Secondary structure elements (α and β) having hydrogen bonds are characterized by reduced fluctuations as compared to other "non-organized" protein parts. The great diversity of the fluctuation amplitudes denotes the protein heterogeneity and the existence of regions varying in mobility and rigidity. In contrast to the residues inside the protein molecule, the motion of side chains, especially those containing charged groups, is strongly affected by the environment (solvent). The atomic motion in itself has an anisotropic character, when random fluctuations differ greatly in their amplitudes depending on the direction of displacement (sometimes the difference being twofold). The direction of the greatest atomic displacement is determined by peculiarities of large-scale cooperative interactions of this atom and neighboring groups. Such cooperative motions occur during a long time (≥10 picoseconds) as compared to the local displacements (~0.2 picoseconds). The correlation of small fluctuating displacements results in the emergence of large-scale cooperative motions involving a large number of atoms. Small displacements also play the role of a "lubricant", facilitating large-scale displacements when protein domains are already involved during longer ("physiological") times. Namely cooperative motions are important for the protein functioning when directed displacements of its structural elements necessary to achieve active conformation occur.

11

Energy Migration and Electron Transport in Biological Structures

Up to now we have analyzed intramolecular mobility of proteins which is spontaneous and caused by thermal motion. But participation of proteins in metabolic processes and mechanisms of their functional activity are always associated with alteration of their electronic state. It is this factor that induces specific conformational transitions to the new equilibrium with the conformation energy minimum, corresponding to the new electronic state of the macromolecule. Electronic transitions and alteration of the electronic state of biopolymers proceed much faster than the conformational rearrangements evoked by them. That is why in the first approximation, electronic and conformational transitions can be considered separately. Electronic transitions in biopolymers play an independent role in a number of foremost biological processes. Of special importance are migration of electronic excitation energy and electron transport in biological structures. Energy migration of electronic excitation for the first time was discovered in experiments on photodissociation of carbomyoglobin which represents the CO-myoglobin complex. Under the influence of light in the presence of oxygen, CO was split off carbomyoglobin and oximyoglobin was formed:

$$\underset{\text{Carbomyoglobin}}{\text{Protein}—\text{Heme}—\text{CO}} \xrightarrow[O_2]{\text{Light}} \underset{\text{Oximyoglobin}}{\text{Protein}—\text{Heme}—O_2}+\text{CO}.$$

It appeared that the efficiency of this process is the same when light is absorbed by the heme (λ ~ 410 nm) or protein (λ ~280 nm). This testifies to the migration of the excitation energy from protein to the heme. Later numerous cases of energy transfer were detected between aromatic amino acids in proteins, nucleotide bases and from a protein to the chromophore dye molecules linked to proteins. An example of energy migration of cardinal biological importance may be the transfer of excitation from light-absorbing molecules of pigments (chlorophyll) to the reaction center in photosynthetic membranes.

An electron transport in biostructures proceeds over large distances without a direct contact between the electron donor and accepter. In mitochondria and chloroplasts, it is electron transport that underlies the key energetic processes – respiration and photosynthesis. In these organelles the distances between different prosthetic groups of carriers directly delivering electrons are about 10–15 Å. A remarkable peculiarity is that at separate stages of the electron transport in biostructures an electron can be transferred with high efficiency at low temperatures, including the temperature of liquid nitrogen and liquid helium. Such low-temperature stages of electron transport have been found in photosynthetic reaction centers. Let us analyze the mechanisms of electron transport and energy migration in biopolymers.

Energy Migration. Light quantum absorption in a complex molecule containing π-electrons evokes transitions between electronic levels (S_0, S_i and T) as shown in Fig. 8.1. In any photobiological process, the role of electronic excitation is to provide energy to overcome the activation barrier at the initial stage of the reaction. However, in photosynthesis the free energy of the final products is higher than that of the original ones (CO_2 and H_2O), and therefore in this case light energy is also stored as the energy of chemical bonds of the products of photosynthesis. The physical cause of electronic energy transfer from the excited donor molecule D* to the acceptor molecule A

$$D^* + A \rightarrow D + A^*$$

is conditioned by the nature of their interaction. The transfer occurs because of the Coulomb interaction of electrons in molecules D and A. During the lifetime of the excited state of molecule D*, alternative electromagnetic field

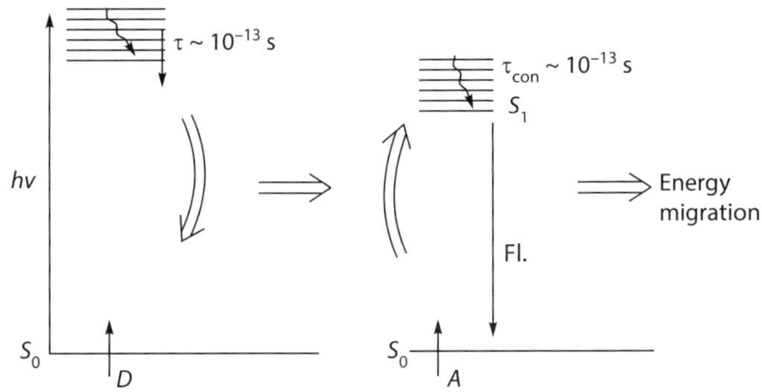

Figure 11.1 Scheme of energy migration of electronic excitation.

is generated due to the oscillations of the electron charge. This field interacts with the electron in the unexcited acceptor molecule A. If the frequency of the alternative field of D* coincides with that of electron transfer in A to an excited level, the excitation energy transfer takes place. As a result the D* molecule returns to the ground state and the A molecule passes to the excited state. No emission of the free light quantum by the D* molecule and its reabsorption by the A molecule takes place so that the transfer is radiationless. The mechanism of energy migration is called inductive resonance and is realized at weak interaction energies between molecules (E ~ 10–3 eV). In this case the time τ_M of energy migration is much larger than 10^{-12} s:

$$\tau_m >> 10^{-12} \text{ s},$$

i.e. it exceeds the time of thermal energy dissipation on vibrational sublevels. This means that upon excitation of each donor molecule it has enough time to pass to the lower vibrational sublevel of the excited state before further excitation transfer or fluorescence emission occurs (Fig. 11.1). Two pendulums tied by a rope are a classical analogue of the inductive resonance mechanism of energy migration (Fig. 11.2). The oscillation energy of one pendulum is transferred mechanically to the other which also begins to oscillate while the oscillations of the first pendulum are decaying. The probability of an inductive resonance transition is proportional to the extent of overlapping of the donor fluorescence spectrum and the acceptor absorption spectrum (Fig. 11.3) and inversely proportional to R^6 where R is the distance between interacting molecules. The distances at which the inductive resonance transfer of energy is efficiently realized are about 20–50 Å at the rates of 10^6–10^{11} s^{-1}. Upon temperature decrease the spectral fluorescence bands D and absorption bands A become narrow hence the

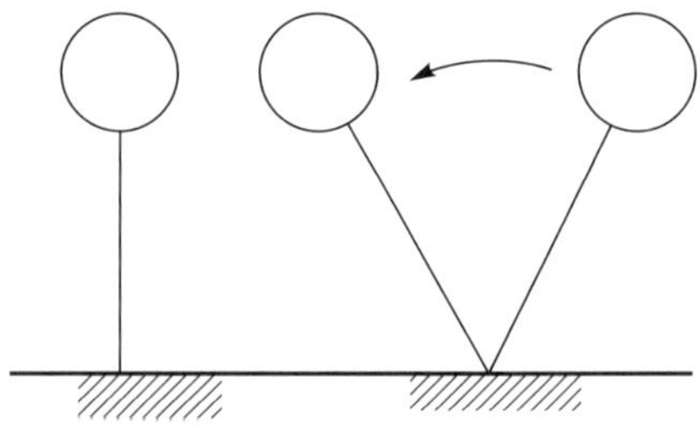

Figure 11.2 Mechanical analog of the inductive resonance mechanism.

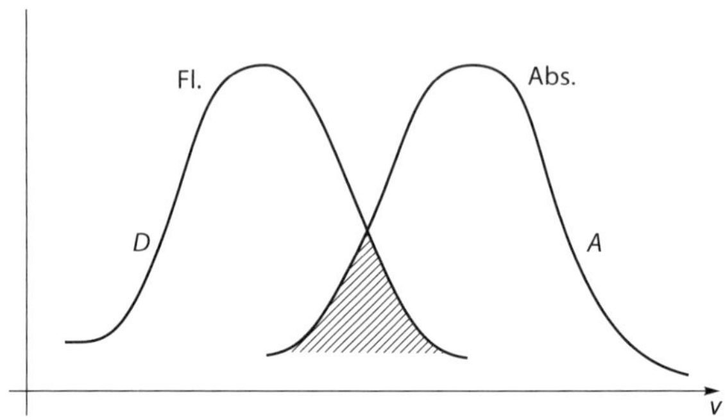

Figure 11.3 Overlapping of the donor fluorescence spectrum (D) and the acceptor absorption spectrum (A).

extent of their spectra overlapping decreases and the probability of energy transfer diminishes.

Thus carotenoid molecules, that absorbed light and turned into state S_1, can transfer the excitation energy to chlorophyll because their S_1 level are higher than the S_1 level of chlorophyll

$$\mathrm{Car}_{S1}\downarrow\uparrow + \mathrm{Chl}_{S0}\downarrow\uparrow \rightarrow \mathrm{Car}_{S0}\uparrow\downarrow + \mathrm{Chl}_{S1}\downarrow\uparrow.$$

The inductive resonance transfer may take place not only between singlet levels of donor and acceptor molecules, but also by the triplet-singlet ($D_T \rightarrow A_{S1}$) and singlet-triplet ($D_S \rightarrow A_{T1}$) mechanisms. At shorter distances of about 1–3 Å (chemical bond length) donor and acceptor electronic orbitals may overlap. Then the excitation transfer proceeds by the so-called exchange resonance mechanism when electrons and electron states are "exchanged". For example, the singlet-triplet transfer can be represented in the following scheme:

$$D_T\uparrow\uparrow + A_{S0}\uparrow\downarrow \rightarrow D_{S0}\uparrow\downarrow + A_T\uparrow\uparrow.$$

In photosynthetic membranes the exchange resonance transfer may occur from chlorophyll in state S_1 to a lower triplet level of carotenoids

$$\mathrm{Chl}_{S1}\uparrow\downarrow + \mathrm{Car}_{S0}\uparrow\downarrow \rightarrow \mathrm{Chl}_{S0}\uparrow\downarrow + \mathrm{Car}_{T}\uparrow\uparrow.$$

As seen, here the total spin is not preserved in the D and A system.

Another so-called exciton mechanism of energy migration is realized at larger interaction energies (~ 10^{-2} eV). In this case the migration time τ_M is much lower than 10^{-12} s making 10^{-13}–10^{-14} s. The excitation that has reached the donor molecule may get over to the neighboring acceptor molecule earlier than relaxation proceeds to the lower vibrational levels of state S_1 of the donor molecule. It looks like the excitation is running over the vibrational levels of interacting molecules, having no time to reach the equilibrium on each of them individually (Fig. 11.4). In this case, excitation covers simultaneously several hundreds of molecules, so that it has a cooperative character. This type of migration is called excitonic, and the excitation region itself including a large number of molecules is called an exciton. In photosynthetic membranes the exciton mechanism takes place upon energy migration within a group of identical pigment molecules, fixed on the same protein carrier. The transfer between different pigment protein complexes proceeds by the inductive resonance mechanism.

The Tunneling Mechanism provides for an efficient electron transport between donor–acceptor groups positioned at a distance of 10–15 Å. Just such transfer can take place in the photosynthetic and respiratory chains, where prosthetic groups are immersed in protein globules at 5–10 Å and interact with each other via a protein body (in cytochromes). An electron transport occurs in protein along an "electronic path". Let us consider the nature of these processes. The experiments have shown that an electron

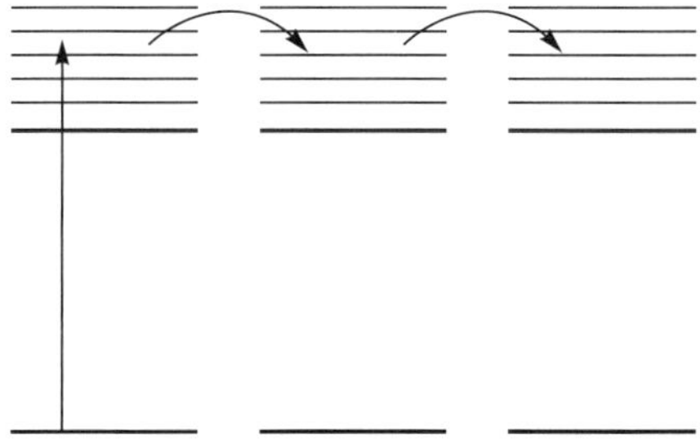

Figure 11.4. Scheme of exiton mechanism of energy migration.

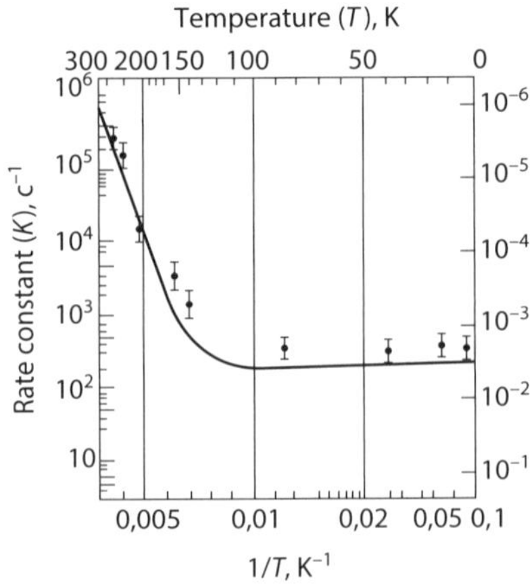

Figure 11.5 Temperature dependence of the rate of cytochrome oxidation in photosynthetic bacteria (Devault D. Quantum mechanical tunneling in biological systems.// QRevBiophys. 1980).

transfer in a photosynthetic chain is efficient both at room and low temperatures. Figure 11.5 shows the dependence of the oxidation rate of cytochrome by a photoactive bacteriochlorophyll molecule in photosynthetic reaction centers.

As seen, the curve has a two-phase character. The initial activation portion of the curve corresponds to the temperature effect on the rearrangements of atomic nuclei in protein parts of carriers required for the effective electron transport. At lower temperatures these rearrangements are hindered and as a result the electron transport rate drops. However even in this case though the transport proceeds slower, at the same time it depends little on temperature. This effect corresponds exactly to the activationless low-temperature part of the electron transport curve (Fig. 11.5). The base of the described electron transport phenomena associated with the rearrangement of the nuclear system is the so-called tunneling effects coupled to electron-conformational interactions in macromolecules. The physical nature of the tunneling effect has a completely quantum-mechanical character and has no classical analogs. Taking into account the importance of electron-conformational interactions comprising the basis of macromolecule functioning, let us consider these problems in more detail.

In accordance with the quantum concepts, a particle (an electron and separate nuclei) has a certain probability to transfer through a potential barrier, the energy of which is higher than the energy of the particle itself

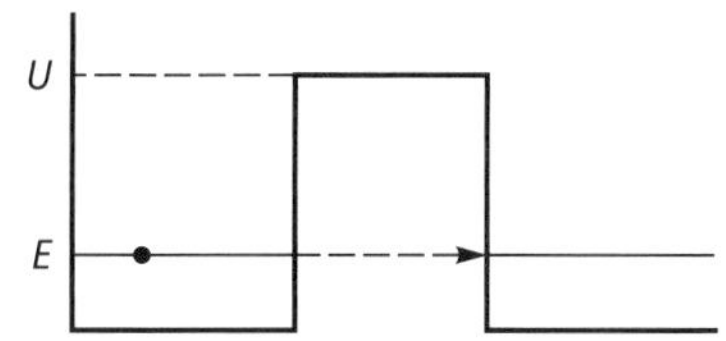

Figure 11.6 Electron tunneling through the potential barrier.

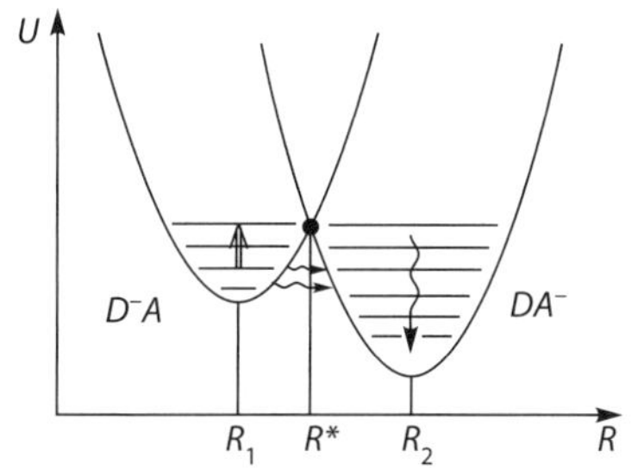

Figure 11.7 Electron–oscillation interactions upon electron tunneling: $R_1 \rightarrow R_2$ is the shift of the equilibrium position of nuclei upon transition from electronic state D^-A to state DA^-.

(Fig. 11.6). Such "percolation" through the barrier or tunneling does not require thermal activation. In quantum mechanics it is associated with some "smearing" of the state of the particle. Consequently, it is possible to find a particle in different sites of its surrounding, including even the region behind the potential barrier. Tunneling transitions are inherent to electrons and nuclei in the DA complex. In the original state (D^-A) nuclear configurations of the donor-acceptor complex are consistent with the state when the electron is localized on the donor (D^-A). After the electron transfer and corresponding change in the electronic state ($D^-A \rightarrow DA^-$) occurs, the nuclear configuration of the final state becomes different from that of the initial state (Fig. 11.7). This means that equilibrium nuclear coordinates R_1 and R_2 of the original (D^-A) and final (DA^-) states of the (DA) complex differ. But there is point R^* where potential energy curves intersect. Evidently at this point the energies of the original (D^-A) and final (DA^-) states coincide. Let us assume that the donor–acceptor complex in state D^-A was rearranged in such a manner that its nuclear coordinate was found in the vicinity of point R^*. This in itself would not necessarily result in the electron transfer. However at points close to R^* the width of the tunneling barrier, separating the potential original and final states, is relatively small. As the energies of the original and final states are similar near point R^*, an electron may have time to tunnel from D^- to A while the system is still near this point. To be fixed on the acceptor, the electron should lose part of its energy and hence not to return back in the same manner. In its turn to achieve this, the nuclear system should be rearranged in such a way that part of the electron energy would be transformed into heat, and the whole system would acquire a nuclear configuration consistent with state DA^- at the R_2 coordinate. In this case, the nuclear configuration changes so that the system would "fall down" to point R_2. As a result an irreversible electron transfer from D to A will take place and hence the system turns into final state DA^-.

The nuclear system of complex DA in the initial state R_1 can reach the vicinity of point R^* because of thermal activation and transition to the upper vibrational levels of the initial state, where nuclear coordinates are close to R^*. This process is consistent with the activation temperature-dependent region of the two-phase curve for the electron transfer. At low temperatures the nuclei are at lower vibrational levels, where the width of the barrier between potential curves for the original and final states is larger than that at the upper levels. In this case the probability of nuclei tunneling to the final state is lower and does not depend on temperature any longer.

So the general probability W of the electron tunneling, associated with the rearrangement of the nuclear system and partial thermal dissipation of electron energy, is summed up of two parts in a simplified form:

$$W = W_0 + W_1 e^{-\hbar\omega / k_{ç}T} . \tag{11.1}$$

Here W_0 is the probability of sub-barrier temperature-independent tunneling from lower vibrational levels; W_1 is the probability of an over-barrier activation transfer process; and $W_0 << W_{1,}$ Þ ω is the energy of vibrational quantum required for the activation.

Electron–conformational Interactions. As seen, the electron tunneling is intrinsic to coupled nuclear system rearrangements. However the latter are characterized by different scales.

We have described the process of initial electron–vibrational interactions (Fig. 11.7) providing for electron tunneling and its fixation in the acceptor molecule due to the partial energy dissipation (< 0.1 eV). The electron fixed on the acceptor induces more profound conformational rearrangements in the complex that makes the base for

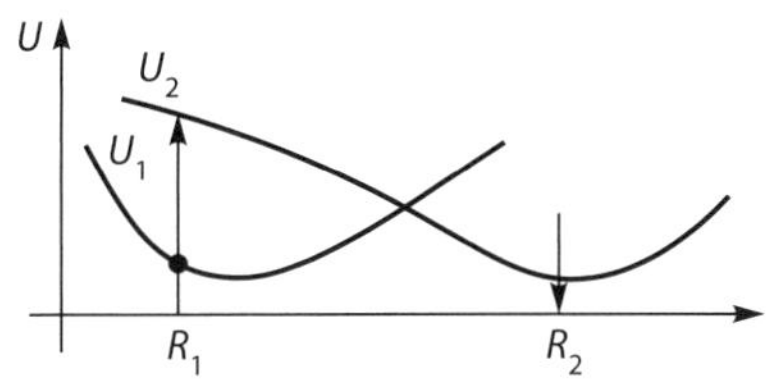

Figure 11.8 Model of a "molecular pump" or electronic–conformational interaction during electron transport. Arrows show "limited diffusion" along conformational coordinate *R*.

electron–conformational interactions. The initial vibrational relaxation occurs at 10^{-2}–10^{-13} s and is coupled to the displacement of nuclei by fractions of angstrom (< 0.1 Å). Conformational rearrangements typically continue much longer (up to 10^{-3}–10^{6} s) and can proceed along with the displacements of nuclei by approximately several angstroms. These rearrangements are already related to the protein functions. Particularly, in a photosynthetic system of electron transport they include the formation of such contact states between the carriers that provide for the directed electron tunneling in the electron transport chain. The coupling of the functional activity of the electron carrier and its intramolecular mobility qualitatively has the following character. In the absence of an electron, prosthetic acceptor group I of the carrier makes stochastic motions by the mechanism of limited diffusion along the conformational coordinate *R* (curve U_1 in Fig. 11.8).

While moving, the group reaches point R_1 where it accepts an electron from the external donor. This step proceeds by the tunneling mechanism with the attachment of the electron and its partial energy dissipation (≤ 0.1 eV) on vibrational degrees of freedom within the donor–acceptor complex. The reduction of group I alters its electronic state and the interaction character with the surrounding within the protein. As a result, it passes from potential curve 1 ($U_1(R)$) to curve 2 ($U_{12}(R)$) of conformational energy. Now, moving stochastically along the conformational coordinate *R*, group I gets to point R_2 where it transfers an electron to the external acceptor and returns back to curve 1 ($U_1(R)$). It should be clearly understood that the stochastic motion in itself along the conformational coordinate is accompanied by energy dissipation and cannot lead to energy accumulation. Following its pathway the molecular group may interact with the environment, for example, with charged "holders" and stay for a long time in certain positions. Thus, a strained conformational state is

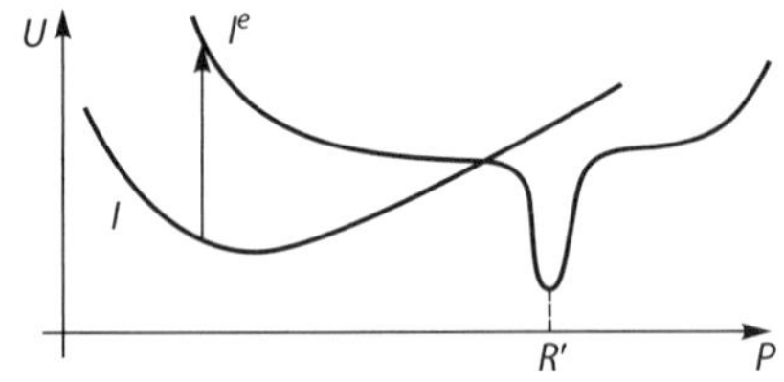

Figure 11.9 Electron transport chain $D \longrightarrow I_1 \longrightarrow I_2 \longrightarrow A$ when "latches" are present in active state R' for reaction $I_1 \longrightarrow I_2$ (tense conformation $\longrightarrow$ sharp minimum) on the conformation potential curve.

created where energy is accumulated. This is consistent with the appearance of a sharp minimum on the conformational potential curve to which the acceptor group has passed after accepting the electron (Fig. 11.9). Point R' may become the location of the positive charge holding group I^e in the strained conformation.

12

Mechanisms of Enzyme Catalysis

Enzymes play a key role in metabolism. They accelerate reactions by increasing their rate constants. Let us analyze the energy profile of a typical reaction (Fig. 12.1) occurring in solutions by the collision mechanism $A + B \rightarrow P$.

Product P is formed if the energy of the colliding molecules of the initial substances A and B exceeds the height of the energy barrier. Apparently it is possible to accelerate this reaction, if the magnitude of the activation energy is reduced.

As known, the generalized scheme of an enzyme reaction includes the formation of a unified substrate complex in the active center where old bonds split and new bonds form thus generating the reaction product. Different theoretical models suggest various ways for decreasing the reaction barrier in the enzyme–substrate complex. As a result of the attachment of the substrate to the enzyme, the reagent entropy is somewhat reduced as compared to their free state. This in itself facilitates further chemical interactions between active groups, strictly oriented relative to each other in the enzyme-substrate complex. It is also presumed that the excess of sorption energy released at the substrate binding is not completely converted into

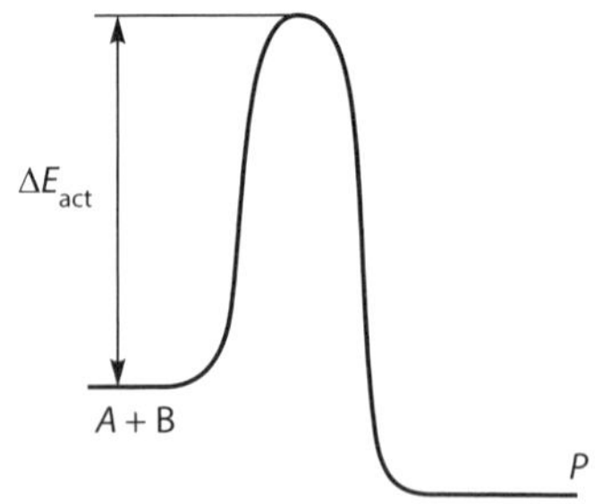

Figure 12.1 Changes in the energy of reagents along the reaction coordinate.

heat. The sorption energy may be partially stored in the protein part of the enzyme and then concentrated on the attacked bond of the substrate in the enzyme–substrate complex. Thereby it is postulated that sorption energy is spent on the formation of the low-entropy energetically strained conformation in the enzyme–substrate complex and hence facilitates the acceleration of the reaction. However experimental attempts to detect such rigid deformations, that could be stored in the protein globule of the enzyme without energy dissipation into heat for a sufficiently long time interval between catalytic events (10^{-2}–10^{-3} s), failed. Moreover, both the mutual orientation required for the catalysis and the drawing together of the splitting substrate bond and active groups in the enzyme center occur spontaneously due to intramolecular mobility of the enzyme and substrate. Such convergence does not involve the formation of energetically unfavorable contacts. This conclusion is inferred from the analysis of non-valence interactions in active centers of a number of enzymes (α-chymotrypsin, lysozyme, ribonuclease and carboxypeptidase). Therefore the strained conformation in the enzyme-substrate complex cannot serve as an indispensable source of energy and the driving force of catalysis.

In other models, it has been presumed that non-dissipative transfer of energy of thermal vibrations from external protein layers to the attacked bond in the active center takes place in the protein globule. However there are no serious arguments in favor of this concept except of the assertion that an enzyme should be constructed in such a way that its structure ensures the coherent character of propagation of conformation fluctuations on certain degrees of freedom without thermal dissipation. In addition to the lack of experimental data, a common disadvantage of these models is that they do not take into account in an explicit form the spontaneous intramolecular mobility of proteins as an important factor. In this respect a step forward was made in the conformation–relaxation conception of enzyme catalysis. Therein the formation of a product is considered as a result of consecutive conformational changes in the enzyme–substrate complex that are induced by the initial alterations in the electronic state of the enzyme active center. At first steps, during a short period (10^{-12}–10^{-13} s) electron–vibrational interactions involving only isolated chemical bonds of the substrate and functional groups of the enzyme proceed not involving the remaining part of the protein globule.

As a consequence, the electronic–conformational nonequilibrium state is formed which relaxes to the new equilibrium coupled to the product formation. The relaxation proceeds slowly and has a directed character, including stages of product release and relaxation of a free enzyme molecule to the initial equilibrium state with the recovery of the enzyme. The coordinate of the enzyme reaction is compatible with the coordinate of the conformational relaxation. As concerns the temperature effect, it affects the conformational mobility rather than the number of active collisions of free reagent molecules, which simply cannot occur in the already formed enzyme–substrate complex. Because of great differences in the rates, we may consider independently both fast short-range electronic interactions, occurring in the active center, and subsequent slower conformational changes in the protein part. At the first stage of catalysis, the stochastic character of the dynamics of the enzyme protein globule and the substrate diffusion to the active center lead to the formation of a strictly definite configuration involving the properly oriented enzyme functional groups and the substrate chemical bonds.

For example, in the case of peptide bond hydrolysis the substrate should be simultaneously attacked by two groups (nucleophilic and electrophilic ones) of the active center so that the reaction could happen. Figure 12.2 demonstrates the relative position of the split peptide bond of the substrate and side chains Ser-195 and His-57. Atom O^{γ}_{155} of residue Ser-195 is at a distance of 2.8 Å opposite carbonyl hydrocarbon C^1, and the proton of the hydroxyl group is at a distance of 2.0 Å above the nitrogen atom of

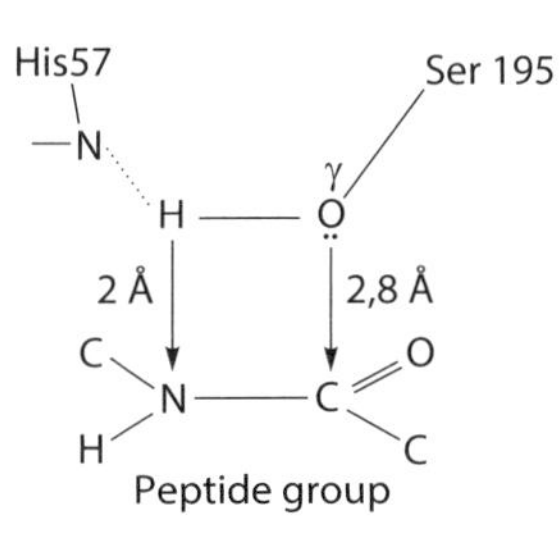

Figure 12.2 Structure of the active center of α-chymotrypsin. The numbers indicate atom-to-atom distances (in angstroms).

the split group not breaking the hydrogen bond with atom N of His-57. The chemical process of catalysis takes place only upon formation of such specific configuration. From the formal point of view, this agrees with a concomitant collision of several molecules, which is very unlikely to occur in a solution.

The question arises: what is the probability of spontaneous formation of such a reactive configuration in a strongly structured medium due to conformational fluctuations of several groups occurring according to the laws of limited diffusion?

Computations show that there is a reasonably definite probability that several groups would simultaneously reach the "reaction" region of a specific radius where they approach each other very closely. This probability is determined primarily by the diffusion coefficient and the number of degrees of freedom of functional groups "searching" for each other in a limited area. For example, upon peptide bond hydrolysis a favorable orientation should be formed for two groups of the active center relative to specific regions of the substrate. Each group has three degrees of freedom, and taking into account the substrate molecule vibrations the total number of degrees of freedom is $N \sim 6$–7. On the whole, this is typical of enzyme processes. It was found that under usual terms the mean time of formation of an active configuration is $\tau \sim 10^{-2}$–10^{-4} s^{-1}, which is consistent with the time of enzyme turnover in conditions of the substrate saturation. In solution this time for a similar reaction is much greater even at high diffusion coefficients. The reason is that having got to the limited region in the strongly structured medium, the functional groups "find" each other and approach each other at close distances earlier than they would "scatter" as it occurs in solution. At the same time, the value $\tau \sim 10^{2}$–10^{-4} s is much higher than relaxation times of separate groups, which is a result of rather rigid steric conditions for the reaction to proceed. The increase in the number of functional groups and formation of indispensable simultaneous contacts between them rises the time necessary to achieve the multicenter active configuration. The overall rate of enzyme catalysis is determined specifically by the time of formation of the required configuration where corresponding groups in the active center come together. The subsequent electronic interactions in such configuration occur much faster and do not limit the overall rate of catalysis.

A number of other peculiarities of enzymes facilitate the substrate transformation in the active center. Thus, as a rule, the microenvironment of the active center with its amino acid residues is more hydrophobic than the aqueous environment. This decreases the dielectric constant of the active center ($\varepsilon < 10$) as compared to that of water ($\varepsilon \sim 80$) and enhances the electrostatic interactions in the hydrophobic medium between the substrate and polar groups of the enzyme. In addition, the protein medium with polarity lower compared to water, partly shields the transferred charges from the action of the polar solvent. And high local concentration of peptide bond dipoles generates electric fields in the active center with the voltage of about thousands and hundreds of thousands of V/cm. So, oriented polar groups generate an electric field affecting the Coulomb interactions in the active center.

To be decoded, mechanisms of electron transitions in the active configuration require the use of quantum chemistry methods. Overlapping of electron orbitals can lead to redistribution of electron density, the generation of an additional charge on the antibinding orbital of the attacked bond in the substrate and its attenuation. It is these processes that happen upon the peptide bond hydrolysis in the tetrahedral complex (Fig. 12.2). Electron density drains down from O^{γ}_{195}–Ser–195 to the antibinding orbital in the peptide bond via interactions of the lone-electron pair O^{γ}_{195}–5 with π-electrons of the C^1 atom of the peptide bond. As it takes place the lone-electron pair of nitrogen in the amine group is expelled from the peptide bond $N{=}{=}C^1$ that loses its dual nature and as a result becomes weak. Concurrently the drainage of the electron density from O^{γ}_{195} weakens the H–O^{γ}_{195} bonding too. In its turn it facilitates the interaction of the H atom in the enzyme and the N atom in the amine group as well as its protonation upon the proton transfer from O^{γ}_{195} to His-57. For a second time it enhances the interaction of O^{γ}_{195} with the peptide group, and so on. Therefore a unique situation is created in the tetrahedral complex when several reactions proceed simultaneously mutually accelerating each other. Concurrent transfer of the charge and proton between Ser-195, His-57 and the peptide bond provides for high efficiency of the process. The catalytic process combines three separate bimolecular reactions, that lead to the disruption of the peptide bond (an unlikely event in solution), in a concert cooperative system. In this system, natural conformational rearrangements are induced and as a result deacylation of the enzyme and protonation of the O^{γ}_{195} atom take place. The principle of formation of a polyfunctional closed system of atomic groups in the active configuration is also realized in other enzyme–substrate complexes (Fig. 12.3).

In enzyme catalysis, the multistage character of substrate transformations, which is unlikely to occur in a solution, is provided due to the concurrent cooperative occurrence in the concert polyfunctional system. Substitution of coordinated processes for ineffective activation stages leads

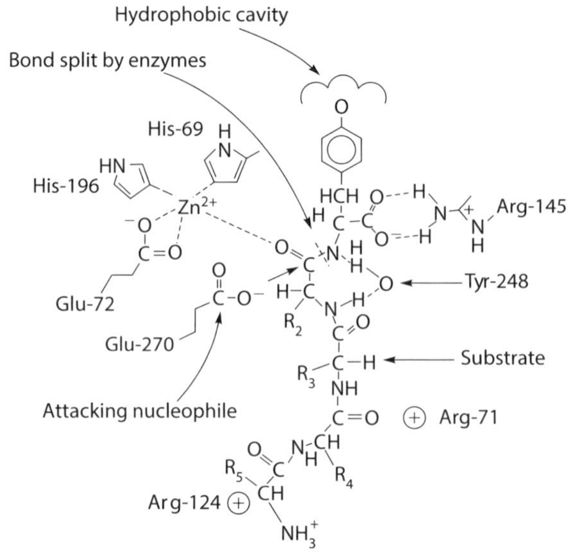

Figure 12.3 Formation of an active configuration or chain of redistribution of bonds upon peptide bond hydrolysis with carboxylase A.

to the reduction of the activation energy of the overall reaction. Let us remind once again that strictly speaking, the "activation energy" concept in enzyme catalysis does not have the same physical meaning as compared to reactions in solutions proceeding by the mechanism of active collisions of free molecules.

13

Physicochemical Features of Biological Membranes. Ionic Equilibria

Biological membranes are mainly composed of proteins, lipids, and hydrocarbons. The composition of a natural lipid molecule comprises a polar charged phosphate head and long hydrocarbon chains of fatty acids. In natural phospholipids, fatty acids may have saturated double bonds mainly in position 2 of the glycerine residue. Proteins may run through the membrane and may be partly or entirely submerged into the lipid layer. The interaction with hydrophobic lipids is implemented mostly by nonpolar amino acid residues. Proteins are floating in the lipid layer of the membrane as isolated globular particles and possess definite mobility. The activity of membrane proteins depends on both the phase state of lipids and the membrane viscosity. Figure 13.1 shows a common scheme of the membrane structure consisting of a lipid bilayer with protein molecules submersed in it. The thickness of biological membranes does not typically exceed 100 Å.

It is convenient to study physicochemical properties of biological membranes using models of monolayers obtained by applying lipids on the water surface. The increased pressure leading to higher compactness of the monolayer reduces the mobility of hydrocarbon chains and enhances their

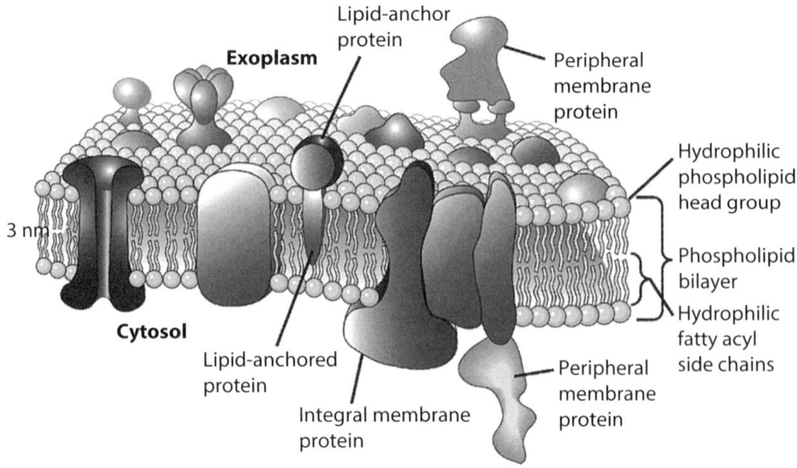

Figure 13.1 Evolution of the conception of the molecular structure of biological membranes (G.L.Nicolson, 2013).

interaction with each other, exposing polar heads to the surface of the phase interface. In the limit, the monolayer becomes so compact that the area of the lipid molecule cross-section is independent of the hydrocarbon chain length. Since the monolayer represents only a half of the membrane lipid bilayer, various artificial bilayer lipid membranes (BLM) prove to be more suitable models. Two-dimensional lamellar structures may fuse forming closed vesicular particles (liposomes) in which lipid bilayers separate the internal aqueous phase from the external solution. Protein molecules and other components of biological membranes can be inserted in vesicular particles in order to investigate the mechanisms of their functioning in biomembranes. Two-dimensional BLM are used to study the bilayer function of electromechanical characteristics and intermolecular interactions in membranes. Electrostatic interactions occur between charged groups either within one half-layer (lateral interactions) or between different layers (transmembrane interactions). Van-der-Waals dispersion interactions between membrane surfaces are found at the distances up to 1000 Å. This is much larger than the distances where electrostatic repulsion is manifested. The total effect of these forces can lead to the appearance of the minimum of the interaction energy at distances from 30 to 80 Å and fusion of cell membrane surfaces. This effect underlies the association of individual cells into cell aggregates. The components of cell membranes are characterized by specific mobility. The characteristic time of rotational motion τ_{rot} of phospholipid and fatty acid molecules in natural membranes is about 10^{-9}

s and rises to 10^{-8} s at temperatures below the melting point of fatty acid chains in lipids. Lateral diffusion of lipids along the layer is characterized by a rather high diffusion coefficient $D \sim 10^{-7}–10^{-8}$ cm²/s, the magnitude of which is strongly dependent on the composition of membranes and temperature (E_{act} of diffusion is ~10 kcal/mol). Transmembrane transitions of lipids from one layer to the other (flip-flop transitions) are much slower ($\tau_{transmembr}$ ~ 1000 s). The mobility of protein molecules in membranes is far lower (τ_{rot} ~ $10^{-4}–10^{-6}$ s, $D \sim 10^{-10}–10^{-12}$ cm²/s). The viscosity of the hydrocarbon region in membranes is typically 1–2 poises, which suggests its liquid-phase state, though this value is two orders of magnitude higher than the viscosity of water. The strength of membranes is determined by their chemical composition and external conditions and can be impaired due to local mechanical defects caused by compression. The average lifetime of BLM in the electric field drops with an increase in the membrane voltage (the electrical breakdown of the membrane). The membrane energy depends on the surface tension, i.e. on the work that should be applied to form one square centimeter of the lipid layer surface. If the membrane has a defect like a through pore, the membrane energy becomes dependent on the pore radius. The energy changes by a magnitude equal to the total surface tension on the area occupied by the pore and also due to the change in the electric capacitance of the membrane caused by the structural defect. As a result, the dependency of the membrane energy on the defect radius has a maximum (Fig. 13.2). It is seen that defects of the small radius disappear, but defects of the radius exceeding the critical one (φ_0)

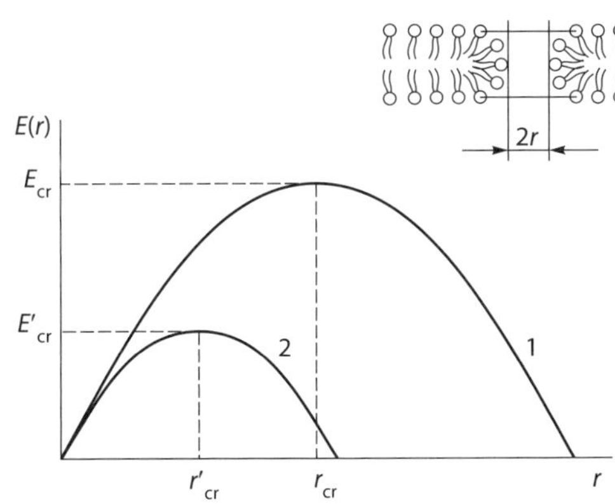

Figure 13.2 Dependences of the defect energy in membrane $E(r)$ versus defect radius r in the absence and presence of electric field (1, at membrane potential $\varphi = 0$; 2, at $\varphi > 0$). Top insert on the right shows representation of a pore in the lipid bilayer (по Антонову В.Ф., 1975).

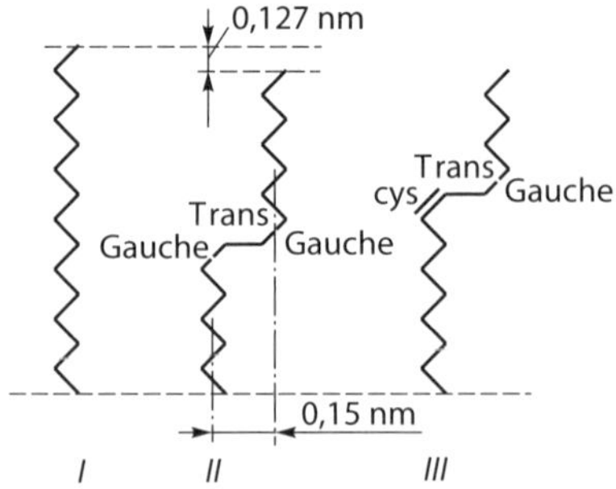

Figure 13.3 Schematic representation of hydrocarbon chains in complete transconfiguration (III). Kink blocks in hydrocarbon chains of membranes.

irreversibly increase resulting in the membrane disruption. The critical radius decreases when the potential difference is applied to the membrane, which explains the increased probability of mechanical disruptions of membranes in the electric field.

Physicochemical properties of membranes are strongly dependent on the lipid phase which is changed upon reaching the critical temperature of phase transition. The phase transition temperature rises with the increasing chain length and drops with a larger number of double bonds in fatty acid residues. As a rule, in natural conditions the major part of natural lipids with unsaturated bonds is in a "liquid" state. The point of phase transition for such lipids is attained below 0°C. In a gel-like state, hydrocarbon chains are in a transconformation, and in a liquid-like state they are disordered. Upon melting the hydrocarbon chain is partially twisted forming a loop or a kink (Fig. 13.3). The increased number of kinks contributes to the disordering of hydrocarbon regions. The kink may shift along the hydrocarbon chain due to a synchronous rotation by 120° of the corresponding sequence of the C–C-bonds. Such a shift of the kink represents atypical diffusion of the free volume within which small molecules may be transferred across the hydrocarbon region.

Lipid Peroxidation. Lipid peroxidation, the process observed upon regular functioning of biomembranes, is usually intensified at pathological states of organisms.

An important role in studying the mechanism of this phenomenon belongs to the Bach-Engler theory of peroxide oxidation (1897), the Semenov theory of branched chain reactions (1934), as well as to the works of Prof. N.M.Emanuel's school on liquid-phase oxidation of hydrocarbons and the works of Prof. B.N.Tarusov's school on the function of chain free radical oxidation of lipids in cell damage.

The process of lipid peroxidation begins with the formation of free radical products as a result of disruption of the C–H- or C–C-bonds in the hydrocarbon molecule. It occurs more readily in hydrocarbons with unsaturated double bonds. The least disruption energy is inherent to the C–H-bond that is adjacent to the double bond in the α-position. But even in this case it is required to use an impact more than 300 kJ/mol to break the bond. That is why the overall rate of lipid peroxidation is determined primarily by the stage of initiation of the free radical process.

Lipid peroxide ROOH is formed in the reaction of the free radical intermediate, namely alkyl radical $R^{\cdot}$ with oxygen. First peroxide radical $ROO^{\cdot}$ is formed, and then after its interaction with substrate molecule RH the following peroxide emerges:

$$R^{\cdot} + O_2 \rightarrow ROO^{\cdot} + RH \rightarrow ROOH + R^{\cdot}.$$

Reactions, in the course of which the original hydrocarbon radical R is regenerated from the nonoxidized molecule of substrate RH, are called chain reactions. The number of cycles that the free radical center manages to complete prior to destruction regulates the chain length.

It has been found that intact biomembranes have enzyme and non-enzyme systems capable of initiating lipid peroxidation. Activated oxygen in the form of superoxide anion radical $O_2^{\cdot-}$, peroxide radical $HO^{\cdot}_2$, hydroxyl radical $HO^{\cdot}$, and singlet oxygen $^1O_2$ initiates free radical oxidation.

Hydrogen peroxide plays a key role in initiating oxidative reactions. Consecutive single-electron reduction of oxygen to water causes the formation of active intermediate products.

At the first stage of oxygen reduction, superoxide anion radical $O_2^{\cdot-}$ is formed. Its oxidation–reduction potential is −0.32 V, which is close to the potentials of reduced pyridine nucleotides. In proton-rich solvents $O_2^{\cdot-}$ is protonated to perhydroxyl radical $HO^{\cdot}_2$.

Further reduction of superoxide anion radical results in the formation of hydrogen peroxide. In the presence of ions with variable valency this compound disintegrates to hydroxyl radicals $HO^{\cdot}$ in an alkaline medium. At the next stage of single-electron reduction of hydrogen peroxide, the bond between oxygen atoms is broken and a water molecule and hydroxyl radical $HO^{\cdot}$are formed. The hydroxyl radical has an extremely high reactivity.

The formation of activated oxygen in the cell proceeds during aerobic respiration and photosynthesis. As a rule, the efficiency of this event enhances upon "overloading" of the electron-transport chain when electron carriers become reduced.

Besides, oxygen activation may take place in the system of endoplasmic reticulum membranes, where electron transport proceeds from reduced pyridine nucleotides via the electron carrier cytochrome P-450. As it takes place, steroid hormones, cholesterol, bile acids and xenobiotics (lipophilic alien molecules) are hydroxylated similar to monooxigenase reactions:

$$S + DH_2 + O_2 \rightarrow SOH + D + H_2O_2,$$

where S is the oxidation substrate and DH_2 is the donor of NADPH electrons.

A specific system of activated oxygen generation functions in phagocytic cells as well.

Activated oxygen is involved in biosynthetic processes, for example, upon the formation of prostaglandins in animals or the phytohormone ethylene in plants.

In photosynthetic reactions, the energy may migrate from the excited pigment molecule (chlorophyll) to oxygen to generate singlet oxygen ^1O. This process may cause destruction of photosynthetic membranes enriched with unsaturated lipids. This also underlies the photodynamic effect of light, when colored molecules serve as energy donors for the O_2 molecule. The singlet oxygen generated in photochemical reactions is to a great extent responsible for the bactericidal effect of sunlight.

It is presumed that both in physiological conditions and under oxidative stress caused by various factors, reactive oxygen species can initiate endonuclease-induced DNA fragmentation (apoptosis and programmed cell death) or even exert a direct effect on DNA (hydroxyl radical).

Reactive oxygen species (ROS) are supposed to play the role of a messenger in regulation of cell homeostasis. Unsaturated lipids of biological membranes represent one of the most susceptible targets for activated oxygen.

The process of peroxidation involves mainly unsaturated phospholipids of biological membranes, primarily polyunsaturated phosphatidylethanolamine. Peroxidation causes the enhancement of the membrane viscosity due to the reduction of the number of liquid hydrophobic lipids in bilayer regions, the formation of intermolecular cross-links and the increase in the number of ordered lipids with constrained mobility in the bilayer. The negative charge on the membrane surface increases as well, which is initiated by the emergence of secondary products of peroxidation of lipids, containing carbonyl and carboxyl groups.

In biological membranes, peroxidation induces permeability to various ions, nonelectrolytes and macromolecules. This effect of the loss of the membrane barrier by membranes underlies the pathogenesis of many diseases.

Biological systems are adapted to aerobic conditions. They are protected from toxic effects of ROS by enzyme systems, antioxidants and singlet oxygen quenchers.

The concentration of hydrogen peroxide in the cell is regulated by catalase, which decomposes peroxide without formation of active products, and by peroxidases, which utilize peroxide for oxidation of diverse substrates.

The most typical antioxidants are phenolic compounds capable to interact efficiently with active lipid free radicals with the formation of inhibitor radicals of low activity. In various natural membranes these are tocopherols and a number of hydroquinones.

Another group of antioxidants interacts proficiently with ROS drastically reducing the concentration of superoxide radicals in the system and catalyzing the dismutation reaction (superoxide dismutase or SOD) with the formation of peroxide and oxygen in the triplet state:

$$2O_2^{\cdot -} + 2H^+ \rightarrow H_2O_2 + O_2.$$

The rate constant of this reaction ($2 \cdot 10^9$ $M^{-1}s^{-1}$) is very high.

Quenchers of singlet oxygen (β-carotene and α-tocopherol) play the role of antioxidants. The antioxidant effect has the mechanism of physical quenching when the singlet oxygen state is deactivated. The process of chemical quenching of $^1O_2$ is accompanied by the inhibitor oxidation and, as a rule, occurs at a lower rate.

On the whole, in biological systems adapted to aerobic conditions, the toxic action of ROS is inhibited to a greater or lesser extent by enzyme systems, antioxidants and singlet oxygen quenchers. The content of superoxide radicals in the cells is controlled by superoxide dismutase which is the main cause for elimination of the superoxide anion.

It is likely that chemiluminescence observed in biological systems proceeds with involvement of activated oxygen molecules mostly as a result of decay of products of oxidation of organic compounds, but not due to the direct quantum emission by excited singlet oxygen.

Usually normal activity of animal and plant tissues is accompanied by spontaneous chemiluminescence. Such luminescence is generally enhanced under stress, which testifies to the activation of oxidation processes in these conditions. The Russian school of Prof. B.N.Tarusov has contributed greatly to studying the nature of luminescence in living organisms.

Chemiluminescence is one of the most sensitive methods for determining and evaluating the level of reaction processes involving ROS in biological systems. In many cases in order to detect and determine the concentration of activated oxygen, a chemiluminescent indicator – luminol or lucigenin – is added to the system studied. These compounds are oxidized by ROS and at the same time emit chemiluminescence with a high quantum yield.

Free radicals $R\cdot$ and $RO\cdot_2$ formed during peroxidation can recombine with each other releasing a large portion of energy (~ 70–100 kcal/mol). This is sufficient for the transition of reaction products to the excited electronic state. The subsequent transition of excited products to the ground state can be accompanied by the emission of chemiluminescence quanta.

Ionic Equilibria. The difference of chemical potentials (μ) of the substance distributed between membrane regions is the driving force for the diffusion across the membrane. The chemical potential is defined as:

$$\mu = \mu_0 + RT \ln C, \tag{13.1}$$

where μ_0 is the standard chemical potential, and C is the substance concentration. If transfer of ions does occur, their motion depends not only on the concentration but also on the electric potential (φ). In this case the notion of electrochemical potential ($\bar{\mu}$) is used:

$$\bar{\mu} = \bar{\mu}_0 + RT \ln C + ZF\varphi, \tag{13.2}$$

where Z is the ion valency, and F is the Faraday constant (96500 C/mol). The direct driving force of ionic transfer is the gradient in the electrochemical potential $d\bar{\mu}/dx$. The main energy barrier that hinders ion penetration into the lipid membrane from the aqueous phase emerges because the dielectric constant of lipids $\varepsilon \sim 2–3$, while in water it is about 81. As a consequence, the ion energy as a charged particle of radius r and charge e equal to

$$E = e^2/2\varepsilon r, \tag{13.3}$$

increases in the lipid layer. Hence the distribution coefficient of ion concentrations between the aqueous and lipid phases should be very low:

$$\gamma = \frac{C^{\text{lip}}}{C^{\text{H}_2\text{O}}} \sim 10^{-20} - 10^{-30}, \tag{13.4}$$

where C^{lip} and $C^{\text{H}_2\text{O}}$ are ion concentrations in phases at the interface.

Let us analyze the distribution of electrolyte A^+B^- in which anion B^- and cation A^+ have different lipophilicites (given lipophilicity of anion B^- is higher) between lipid (2) and aqueous (1) phases. Near the interface the concentration of anions B^- in the lipid phase would be somewhat higher than that of cations A^+. Consequently, near the interface the nonpolar lipid phase (2) carries a net negative charge and therefore has a lower potential compared to the aqueous phase (1). However in phase regions far from the interface surface, due to electrical neutrality of the bulk phase volume A^+ and B^- concentrations are equal and $C_A = C_B$. On the whole, the concentration of both types of ions (A^+ and B^-) is lower in phase (2) due to the difference in dielectric constants of phases I and II.

Figure 13.4 shows potential profiles of A^+ and B^- concentrations and the electric potential profile near the phase boundary. As seen, the regions near

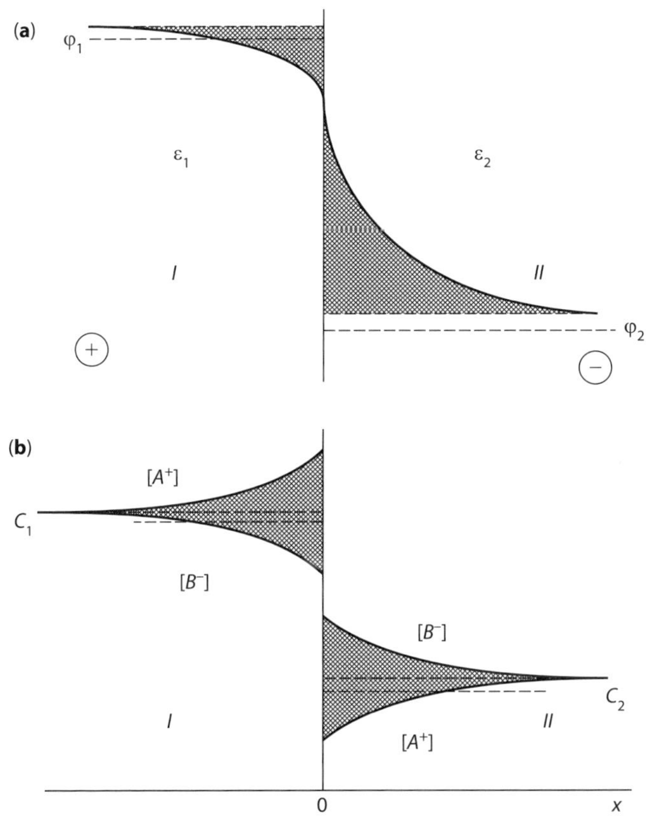

Figure 13.4 Profiles of the electric potentials and concentrations of A^+ and B^- ions on the interface of phases I and II. Phase I, aqueous solution ($\varepsilon_1 \sim 80$); phase II, nonpolar solvent($\varepsilon_2 = 2 \div 3$). a, Potential profile (φ); b, concentration profiles (C) of cations and anions. C_1 and C_2 are electrolyte concentrations in the volume of phases I and II.

the interface have diffusion layers, where A^+ and B^- concentrations are dissimilar so that the condition of electrical neutrality is not valid near the interface. The proportion of the electrolyte concentrations in the two phases can be determined from the equilibrium condition at the phase interface

$$\mu_{OA}^{(1)} + RT\ln C_A^{(1)} = \mu_{OA}^{(2)} + RT\,\ln C_A^{(2)}. \tag{13.5}$$

Here terms $F\varphi_1$ and $F\varphi_2$ are absent because the potentials of each phase are equal at the interface $\varphi_1{}^0 = \varphi_2{}^0 = \varphi_0$. Hence taking into account expression (13.4) $\gamma_A = C_A^{(2)} / C_A^{(1)}$ we obtain that for γ_A and γ_B

$$\ln \gamma_{A+} = \frac{\Delta\mu_{OA}}{RT}, \quad \ln \gamma_{B-} = \frac{\Delta\mu_{OB}}{RT}, \tag{13.6}$$

where $\Delta\mu_{OA} = \mu_{OA}^{(1)} - \mu_{OA}^{(2)}$, $\Delta\mu_{OB} = \mu_{OB}^{(1)} - \mu_{OB}^{(2)}$.

Given electrical neutrality of each phase $C_B^{(1)} = C_1$ and $C_A^{(2)} = C_B^{(2)} = C_2$,

$$\ln.\frac{C_2}{C_1} = \frac{1}{2}(\ln \gamma_A{}^+ + \ln \gamma_{B-})$$

or

$$\frac{C_2}{C_1} = \sqrt{\gamma_{A+}\gamma_{B^-}}. \tag{13.7}$$

The relation of electrolyte concentrations in the phases depends on γ_{A+} and γ_{B-}. The magnitude of the electric potential difference $\Delta\varphi = \varphi_1 - \varphi_2$ between the phases depends on the difference of electrochemical potentials of both ion species. It can be demonstrated that in conditions of electrical neutrality of each phase

$$\Delta\varphi = \varphi_1 - \varphi_2 = \frac{1}{2F}(\Delta\,\mu_{OB} - \Delta\,\mu_{OA}) = \frac{2F}{RT}(\ln\gamma_A - \ln\gamma_B). \tag{13.8}$$

Thus, it follows from expression (13.8) that the interfacial potential difference is generated only if the distribution coefficients of cations and anions are different ($\gamma_A \neq \gamma_B$). The characteristic of the potential drop near the interface depends on ion distribution in the electric double layer over the depth of the membrane. The analysis shows that relative to the phase volume the value of the electric potential near the interface changes exponentially with the distance along coordinate x

$$\varphi(x) = \varphi_0 e^{-\chi|x|}, \tag{13.9}$$

where φ_0 is the potential at the interface, and χ is the constant depending on both the ion concentration in the given phase and the dielectric constant. The major part of the potential drop (by e times) occurs at the distance $x = 1/\chi$ which characterizes the depth of the electric double layer.

The quantity $\lambda = 1/\chi$ is the screening length. It diminishes with an increase in the electrolyte concentration in the given phase. In diluted solutions, the depth of the double layer makes hundreds of angstroms, while in concentrated solutions it is only units of angstroms. This means that in a lipid layer where ion concentrations are several orders lower than in aqueous medium, the screening length is much longer. In a thin membrane, the thickness of which is far smaller than the screening length, the potential jump within the membrane is almost lacking. In the absence of the external field the potential within a thin membrane remains constant over its thickness (Fig. 13.5). Therefore the electric potential in the thin uncharged membrane does not differ from the potentials of the surrounding solutions at any distance from the membrane. It is just the situation that takes place in uncharged bilayer lipid membranes, the thickness of which makes 70–100 Å. If an external field is applied to such thin uncharged membrane, the potential will change linearly within the membrane, i.e. the gradient of the applied potential in the membrane is constant over its thickness: $d\varphi/dx$

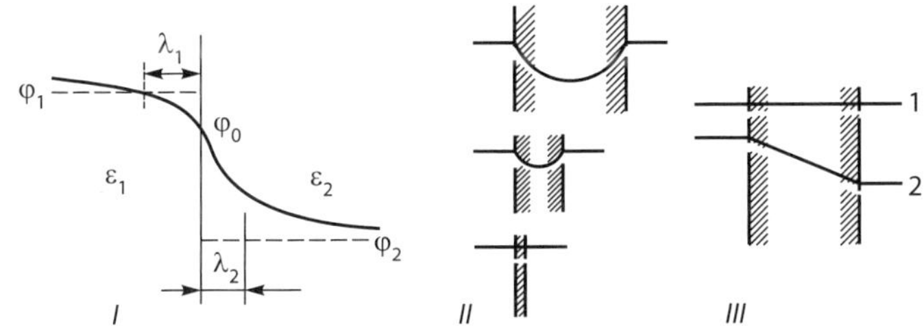

Figure 13.5 Potential profiles on the phase interface (*I*) in membranes of different thickness (II), in a thin membrane in the absence of the external field (III, 1) and upon the application of the electric potential difference (III, 2). φ_1 and φ_2 are electric potentials in the bulk phase; φ_0 is the electric potential on the interface; ε_1 and ε_2 are dielectric constants of the two phases; λ_1 and λ_2 are the Debye screening lengths for the first and second phases, respectively.

= const. But if the surface of the thin membrane is initially charged and has fixed charges, an electric double layer will be formed near the surface. In this layer counterions from the solution will be retained by the electrostatic attraction of the membrane charges. This retention is impeded by thermal motion of ions tending to equalize their concentrations on the surface and in the volume. As a result, the surface potential on the membrane will depend intricately on the density of the surface charge, the equilibrium concentration of electrolyte in the solution and temperature.

The membrane surface potential plays an important role in bioelectrochemical processes. In experiments a related value – electrokinetic potential ξ (zeta potential) or sliding potential – is usually measured. It is determined taking into account that the first layer of ions with their hydration shells and the first layer of water molecule, which are wetting the solid phase, are retained near the charged surface when the membrane particle moves relative to the fluid. Electrokinetic phenomena (electroosmosis, electrophoresis, streaming potential, and sedimentation potential) are determined by the membrane surface charge.

Donnan Equilibrium is a type of ionic equilibrium of phases when one phase has charged particles incapable of permeating to the other phase. Assume that two aqueous phases, with electrolyte A^+B^- dissolved in each of them, are separated by a membrane permeable for the electrolyte. Let positive fixed charges Q that cannot pass across the membrane, be in one of the phases (phase 2). At the equilibrium for a mobile ion species, the electrochemical potentials (13.2) in both solutions (1 and 2) are the same

$$RT\ 1\text{n}C_{A1} + Z\text{F}\varphi_1 = RT\ 1\text{n}C_{A2} + Z\text{F}\varphi_2 \tag{13.10}$$

and

$$RT\ 1\text{n}C_{B1} + Z\text{F}\varphi_1 = RT\ 1\text{n}C_{B2} + Z\text{F}\varphi_2.$$

From conditions of equilibrium (13.10) the following equation for the potential difference is obtained:

$$\varphi_1 - \varphi_2 = \frac{RT}{\text{F}} 1\text{n}\ \frac{C_{A_2}}{C_{A_1}} = -\ \frac{RT}{\text{F}} 1\text{n} \frac{C_{B_2}}{C_{B_1}}, \tag{13.11}$$

which is called the Nernst equation for an equilibrium potential on the membrane. The equation shows that the equilibrium potential on the membrane is determined by the proportion of electrolyte concentrations in both phases. If C_{A1}, = C_{B1}, = C_{A2}= C_{B2}, = C, then $\Delta\varphi = 0$. From this equation we obtain

$$C_{A2} \cdot C_{B2'} = C_{A1} \cdot C_{B1'} = C^2. \tag{13.12}$$

The effect of nondiffusing ion *Q* generates the electric potential difference even if the electrolyte concentrations on both sides of the membrane were initially equal. Indeed, it follows from the electrical neutrality of the first phase that

$$C_{A1} = C_{B1'} = C. \tag{13.13}$$

while for the second phase the same condition in the presence of *Q* looks like:

$$C_{A2} + Q = C_{B2}. \tag{13.14}$$

It is evident that in the second phase $C_{A2} < C_{B2}$, and since according to (13.12 and 13.13) $C_{A2} \cdot C_{B2}$, = C^2_{A1}= C^2, then $C_{A2} > C_{A1}$. This means that charge *Q* in phase 2 is compensated by an increased content of anions B_2^- in it at the lower content of cations A_2^+. In this case we have the Donnan electric potential difference equal to

$$\Delta\varphi_D = \varphi_2 - \varphi_1 = RT\,\ln\frac{C_{A_1}}{C_{A_2}} = \frac{RT}{F}\ln\frac{C}{C_{A_2}} > 0,$$

where $C_{A2} < C$. The value of $\Delta\varphi_D$ can become quite high at $Q >> C$ when anions B_2^- accumulate in phase 2 to concentrations much higher than the electrolyte concentration *C* in the aqueous medium.

14

Passive Transport of Substances Across Membranes

Passive transport of substances across membranes from one phase to the other is caused by driving forces, specifically concentration gradients or potentials existing between the phases.

Nonelectrolyte Transport proceeds by diffusion mechanisms. The substance flux I towards axis x is proportional to the concentration gradient $\frac{dC}{dx}$ (Fick law of diffusion):

$$I = -D\frac{dC}{dx}, \qquad (14.1)$$

where D is the diffusion coefficient (in $cm^2 \cdot s^{-1}$) dependent on temperature and substance mobility u in the medium:

$$D = RTu. \qquad (14.2)$$

If the membrane is thin, the concentration gradient is constant ($\frac{dC}{dx}$ = const) and the driving force is equal to the difference of concentrations C_1 and C_2 in the solutions separated by a membrane

$$I = P(C_1 - C_2),$$

where P is the permeability coefficient of a membrane to the given substance. Permeability depends on coefficients of diffusion D and distribution γ of substance between aqueous and lipid phases and diminishes with an increase in the membrane thickness h

$$P = \frac{D\gamma}{h} = \frac{RTu\gamma}{h}. \tag{14.3}$$

There is specific correlation between the penetration power of substances and their solubility in lipids which is valid for large hydrophobic molecules. However small hydrophobic molecules can penetrate through aqueous pores or travel following diffusion of kinks across hydrocarbon area. Nonelectrolytes penetrate into hydrophobic part of a membrane or into a narrow pore provided they lose their hydration shell. A certain amount of energy is required to "tear off" water dipoles from polar groups of a molecule (–COOH, –OH, and – NH_2). This is the reason for strong temperature dependence of the membrane permeability coefficient for some nonelectrolytes. Another type of passive transport is the facilitated diffusion of amino acids and sugars. It differs from typical diffusion by its high specificity relative to the transported substance, saturation of the diffusion rate with the increase in the substance concentration, and sensitivity to certain inhibitors. This testifies to the similarity of the facilitated diffusion with enzymatic processes and suggests the involvement of carriers in this process. Theoretically, the initial rate of substance transport is described by an equation analogous to the Michaelis-Menten equation:

$$v = \frac{V_{\max} \cdot S_0}{K_M + S_0} \tag{14.4}$$

where S_0 is the substance concentration in the outer solution. The substance concentration in the interior is S_i=0 at the zero time. With the growth of S_0 all carriers become bound with a substrate and hence the diffusion rate does not increase any longer. The driving force of the transport proceeding with the partaking of a carrier is the gradient of chemical or electrochemical potential of the transported substance. Some carriers can concurrently transport both the amino acid and the ion, immobilizing them on two binding sites (for example, in erythrocytes). Transport of glucose in the cardiac muscle proceeds in the same way coupled to the formation of a complex with a carrier, (Carr) Na^+ and glucose (Gl)

$$\text{Carr} + Na^+ \rightleftarrows \text{Carr } Na^+ + \text{Gl} \rightleftarrows (\text{Carr } Na^+ \text{Gl}).$$

The movement of this complex is facilitated due to the interaction of positive Na^+ charge and the electric field. It should be underlined that in passive transport the energy of metabolism is not expended to maintain gradients. Upon equalizing dissipation gradients due to diffusion, the system reaches equilibrium and unfacilitated diffusion is stopped. In contrast to this, in active transport the maintenance of gradients (i.e. driving forces for transport) is caused at the expense of the energy of the coupled metabolic processes.

Ion Transport across an artificial lipid membrane which is considered to be a homogeneous continuous medium proceeds in a continuous regime. In accordance with the Onsager relations of proportionality between fluxes and their driving forces (Chapter 5), flux I of charged ions is proportional to the electric potential gradient ($d\bar{\mu}/dx$) along axis x. It also depends on both ion mobility u and ion concentration C

$$\mathrm{I} = Cu\,(-\frac{d\bar{\mu}}{dx}). \tag{14.5}$$

Substituting the expression for the electrochemical potential $\bar{\mu} = \mu_0 + RT\ \ln C + ZF\ \phi$ in the above equation and taking into account that $\frac{d \ln C}{dx} = \frac{1}{C}\frac{dC}{dx}$, we get

$$I = -uRT\frac{dC}{dx} - uCZF\frac{d\varphi}{dx}. \tag{14.6}$$

Equation (14.6) is called the Nernst-Planck equation of passive transport of ions in the electrochemical potential field. It can be solved under two simplifying conditions. First, it is possible that the condition of electrical neutrality is valid not only for phases divided by the membrane but also for the membrane itself. This is true when concentrations of anions and cations are equal at any plane along axis x ($C_- = C_+$). Therefore the diffusion electrical potential is established between the two electrolyte solutions due to different mobility of cations (u_+) and anions (u_-). The Henderson equation for diffusion potential $\Delta\varphi = \varphi_2 - \varphi_1$ is the following:

$$\Delta\varphi = \frac{u_+ - u_-}{u_+ + u_-} \cdot \frac{RT}{F} \ln \frac{C_2}{C_1}. \tag{14.7}$$

For example, when two NaCl solutions are in contact, the solution with a lower concentration acquires a negative potential relative to the concentrated solution because of higher mobility of the chloride ion $u_{Cl} \approx 1.5\, u_{Na}$. If the ratio of ion concentrations in the solution is about 10 ($C_1/C_2 \sim 10$), the $\Delta\varphi_{dif}$ value may reach ~12 mV in the case of NaCl diffusion.

The Henderson equation, however, is not applicable for the description of the potential established across cell membranes, because the condition of electrical neutrality over the membrane thickness is not fulfilled for thin cell membranes. In this case the condition of a linear change of the potential over the whole membrane thickness $d\varphi/dx$ = const can be valid, which is true just in thin membranes. In this case the Nernst-Planck equation can be solved and thus one can get the dependence of the total passive ion flux J on the potential difference $\Delta\varphi$ across the membrane and on concentrations of this ion C_1 and C_2 in the phase of the membrane itself at its edges

$$I = \vec{i} - \overleftarrow{i} = \frac{ZF\Delta\varphi u}{h} \cdot \frac{C_1 - C_2 \exp(ZF\Delta\varphi / RT)}{1 - \exp(ZF\Delta\varphi / RT)}. \tag{14.8}$$

Two unidirectional fluxes $\vec{i}$ and $\overleftarrow{i}$ make a contribution to the total flux I. The total flux I is composed of two unidirectional fluxes $\vec{i}$ and $\overleftarrow{i}$. Flux $\vec{i}$

$$\vec{i} = \frac{ZF\Delta\varphi u}{h} \cdot \frac{C_1}{1 - \exp(ZF\Delta\varphi / RT)}$$

is a straight unidirectional flux across the membrane from the outer electrolyte solution with concentration C_0. Flux $\overleftarrow{i}$

$$\overleftarrow{i} = \frac{ZF\Delta\phi u}{h} \cdot \frac{C_2 \exp(ZF\Delta\phi / RT)}{1 - \exp(ZF\Delta\phi / RT)}$$

is a unidirectional flux across the membrane from the internal solution with concentration C_i to the outer solution across the membrane.

The ion concentrations (C_1 and C_2) at the edges in the phase of the membrane itself are correspondingly proportional to the ion concentrations in the outer and internal solutions (C_0 and C_i):

$$C_1 = \gamma C_0$$

$$C_2 = \gamma C_i.$$

Then the equation for the passive flux through the membrane is obtained:

$$I = \frac{ZF\Delta\varphi}{RT}\rho\frac{C_0 - C_i\exp(ZF\Delta\varphi / RT)}{1-\exp(ZF\Delta\varphi / RT)} \tag{14.9}$$

where $\rho = uRT\ \gamma/h$ is the permeability coefficient. Expression (14.9) is known as the Goldman equation in the constant-field approximation ($d\varphi/dx$ = const). Unidirectional fluxes $\vec{i}$ and $\overleftarrow{i}$ are independent given each of them determined only by "its own" concentration, i.e. the concentration of the solution is directed from flux. In this case the ratio of independent unidirectional fluxes will be as follows

$$\frac{\vec{i}}{\overleftarrow{i}} = \frac{C_0}{C_i\exp(ZF\Delta\varphi / RT)}. \tag{14.10}$$

The above relation is known as the Ussing-Teorell flux ratio equation for independent unidirectional fluxes. It is used to verify the passive character of ion transport. The Goldman equation predicts nonlinear dependence of net transmembrane ion flux on the potential difference across the membrane. At equilibrium when the total flux equals zero

$$I = \vec{i} - \overleftarrow{i} = 0,$$

it follows from (14.9) that

$$C_0 = C_i\exp(ZF\,\Delta\varphi/RT)$$

and thus the equilibrium value of the potential

$$\Delta\varphi = \frac{RT}{ZF}\ln\frac{C_0}{C_i}$$

is determined by ratio of concentrations in the outer C_0 and internal C_i solutions. If various ions diffuse through the membrane concurrently, their fluxes contribute to the total transport process. At equilibrium, when there is no electric current across the membrane, the sum of individual ion fluxes equals zero. In cell membranes Na^+, K^+ and Cl^- ion transport is of great importance. In this case the condition for the equilibrium is

$$J_{Na} + J_K + J_{Cl} = 0,$$

where each of the fluxes can be expressed by equation (14.9). The summation of expressions for fluxes of Na, K and Cl ions gives

$$\mathrm{P}_K [\mathrm{K}_0] - P_K [\mathrm{K}_i] \exp(F\,\Delta\varphi/RT) + \mathrm{P}_{Na} [\mathrm{Na}_i] \exp(F\,\Delta\varphi/RT) -$$
$$P_{Cl} [\mathrm{Cl}_i] - P_{Cl} [\mathrm{Cl}_o] \exp(F\,\Delta\varphi/RT) = 0.$$

Hence it is possible to derive an expression for equilibrium membrane potential or cell resting potential

$$\Delta\varphi = \frac{\mathrm{RT}}{\mathrm{F}} \frac{P_K[\mathrm{K}_0] + P_{Na}[\mathrm{Na}_0] + P_{Cl}[\mathrm{Cl}_i]}{P_K[\mathrm{K}_i] + P_{Na}[\mathrm{Na}_i] + P_{Cl}[\mathrm{Cl}_0]}. \tag{14.11}$$

Permeability of Na, K and Cl ions is different, and consequently it is exactly the non-uniform distribution of ions on both sides of the membrane that is a key reason for the emergence of the resting potential. The permeability ratio of these ion species is

$$\mathrm{P}_K : \mathrm{P}_{Na} : \mathrm{P}_{Cl} = 1 : 0{,}04 : 0.05,$$

so that the membrane is almost impermeable to Na^+ ions as compared to K^+ ions.

The values of resting potentials in animal cells are usually as low as −70 mV. In intact axons this value is close to the equilibrium potential for K^+ ions. For the sake of simplicity it may be assumed that in these cases the value of transmembrane resting potential is determined mainly by the equality of electrochemical potentials of K^+ ions, their cell equilibrium concentration being much higher than their equilibrium concentration in the outer solution. At the same time, the low cell concentration value of Na^+ ions is far from its level that should be established at the equilibrium.

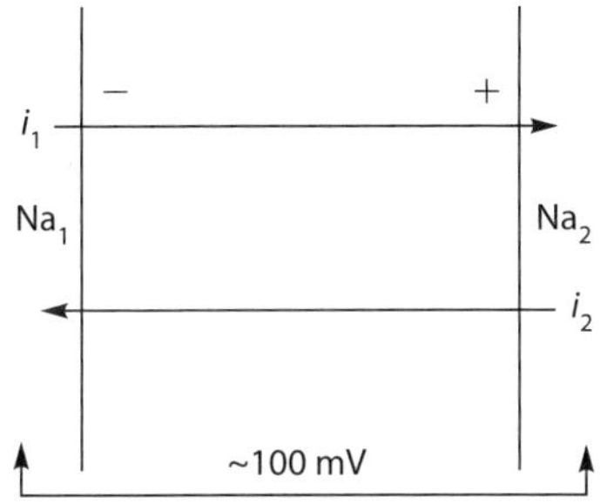

Figure 14.1 Principal scheme for measuring the short circuit current on frog skin.

Hence, to maintain a low non-equilibrium level of Na^+ concentration in the cell, a specific mechanism of active removal of Na^+ ions from a cell by the Na-pump system is evidently required. The experiments on ion transfer through the frog skin suggest the existence of such Na-pump. If two chambers filled with identical Ringer's solution are separated by the frog skin membrane, the potential difference of about 100 mV is formed between its outer and internal surfaces (the outer surface is negative) (Fig. 14.1).

The potential difference can be compensated to zero by applying voltage from an external source. Then the system should reach equilibrium, because now there are no either concentration gradients or potential gradients between the chambers separated by the frog skin membrane. In this case, in agreement with the Ussing equation (14.10), the unidirectional fluxes should be equal: $\vec{i} = \overleftarrow{i}$, at $C_0 = C_i$ and $\Delta\varphi = 0$. But direct experiments with labeled Na^+ showed that in such conditions the Na^+ flux from the outer surface to the internal surface of the skin is much greater than the Na^+ flux in the opposite direction. This process of Na^+ transport from the outer (mucosal) membrane to the internal (serosal) membrane is an active transport. In some plants, the resting potential is as low as −200 mV which greatly exceeds the equilibrium potential of K^+ ions and is explained by active removal of H^+ ions from the cytoplasm.

15

Channels and Carriers. Active Transport

Prior to the analysis of the active transport let us consider mechanisms of ion transfer across membranes.

Channels. A biological membrane has ion channels that are lipoprotein complexes of a complex structure. Narrow channels (a sodium channel of 3.1 × 5.1 E and a potassium channel of 4.5 × 4.5 E) allow for a single-file (one-row) movement of ions which can interact with each other and with molecular groups in the channel. When the ion enters the channel, water molecules of ion hydration shell are substituted for polar groups of the channel cavity. Upon dehydration the increase in the free energy of the ion is compensated by the energy of its interaction with polar groups of the channel. As a consequence, the total energy of the ion diminishes thus facilitating its transport through the channel. The existence of polar groups and fixed anion centers in the channel and their Coulomb interactions with the ion decrease the energy barrier for the ion transport from the solution to the channel. Ions that are strongly bound by electrostatic forces to the anionic centers pass through the channel almost easily. For example, after the loss of its hydration shell the size of Na^+ cation, that is smaller than K^+ cation, will be bound more strongly to the small negative anionic center. However the radius of the hydrated Na^+ ion is larger than that of K^+, and hence, losing its hydration shell Na^+ ion passes worse than K^+ ion through the relatively wide membrane pores. Fixed anionic centers in the channel attract cations and facilitate their transport through the channel by diminishing ion energy. Figures 15.1 and 15.2 demonstrate energy profiles of Na^+ and K^+ channels. The conductance rate of a Na^+ channel reaches 10^7 ions/s. Different organic cations with sizes not exceeding 3×5 Å can also pass through a Na^+ channel. Oxygen atoms from the COO group may play the role of an anionic center. Potassium channels have a wide orifice (>8 Å) from the side of the cytoplasm which may be blocked by tetraethylammonium.

Single-file (One-row) Transport of ions through channels has specific features. An ion stays for a relatively long time in every potential well. This means that another ion cannot enter the occupied potential well because of the electrostatic repulsion with the ion already located in the well. Ion jumping from one well to another well is caused by thermal fluctuations.

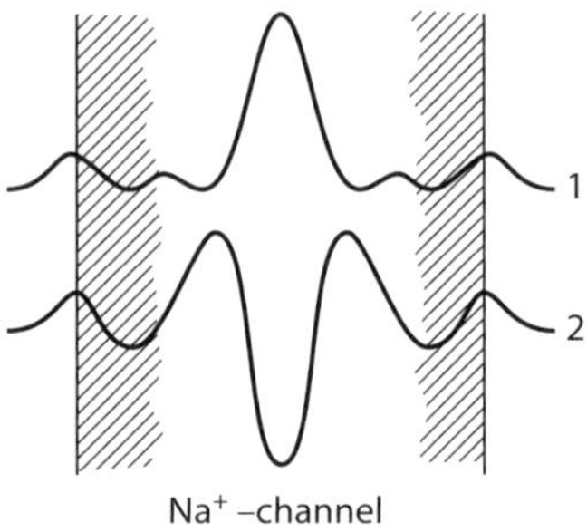

Figure 15.1 Energy profile of a sodium channel in excitable membranes.

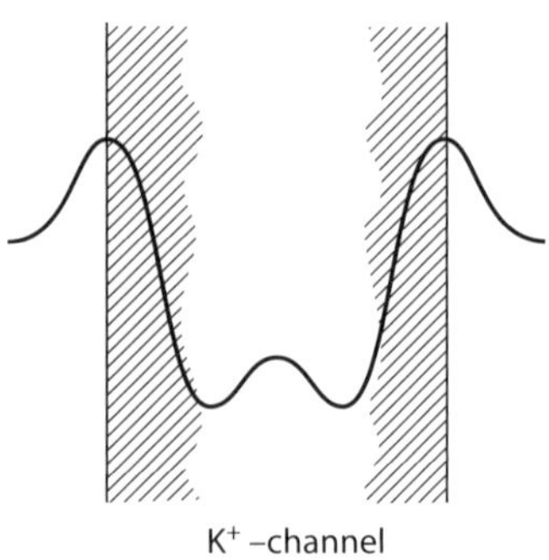

Figure 15.2 Energy profile of a potassium channel in excitable membranes.

The applied external electric filed changes the ion energy and thus affects the jumping probability. The release of the ion bound in an anionic center from the channel exit is facilitated when another ion appears in the channel entrance due to electrostatic repulsion of between the two ions. The channel conductivity depends on to what extent its "inlet" and "outlet" ion binding sites are filled up. At high electrolyte concentrations, the both binding sites may be filled up with ions which would block the channel and saturate its conductivity. It is clear that a change in the number of particles inside the channel alters also its energy profile due to Coulomb interactions. In addition the conformational rearrangements in the protein forming the ionic channel can also change the height of the energy barrier as a result of reorientation of the polar groups.

Carriers. Ion transport across the membrane is also performed via transport of ionophores (carriers). Ionophores can form complexes with ion or form pores (channels) in the membrane. These processes have been studied on bilayer lipid membranes. The energy of the ion–carrier complex is much lower than that of the dehydrated ion. The ionophore–ion complex inside the membrane is formed on one side of the membrane and then moves to the other side where an ion is released and an ionophore is recovered. A typical mobile carrier is valinomycin which mediates K^+ transport. The potassium cation enters the internal cavity of valinomycin, the formed structure being stabilized due to the ion interaction with 6 to 8 polar groups of CO (Fig. 15.3) which replace the hydration shell of the ion. The sodium ion that

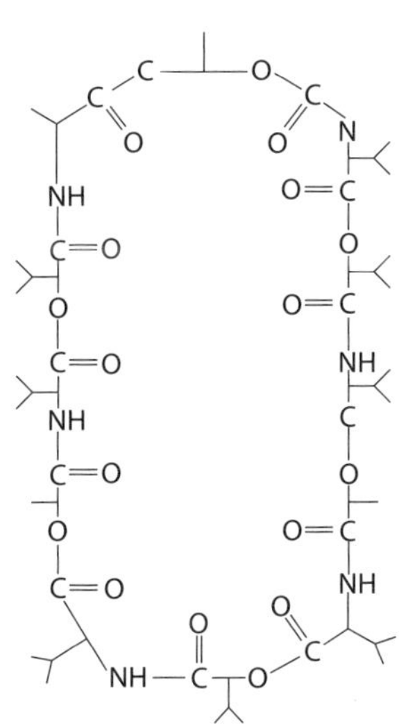

Figure 15.3 Chemical structure of a valinomycin molecule (ionophore).

has a smaller radius is unable to interact effectively with oxygen atoms of the carbonyl groups. A valinomycin molecule transports approximately 10^4 ions per second across a bilayer lipid membrane (BLM).

Another carrier (nigericin) forms complexes with ions, where the molecule is in a folded conformation. Being a weak acid, nigericin can also mediate H^+ ion transport. It induces the exchange of H^+ and K^+ in BLM and biological membranes. An ionophore molecule can form a complex with a hydrophilic pore. The outer part of molecules in such a pore is hydrophobic, and properly polarized groups are directed inwards. Gramicidin is the best known channel-forming ionophore. When in the membrane, a gramicidin *A* molecule formed by fifteen hydrophobic amino acids folds in a helix-like structure. It represents a hollow cylinder of about 30 Å long with the pore diameter of about 5–8 Å. Such a channel may transfer approximately 10^7–10^8 ions per second. When the ion enters the gramicidin channel, part of water in its hydration shell is substituted for carbonyl groups directed inward the pore. Ionic channels exhibit specific fluctuations of permeability caused by their opening and closing, which distinguishes the channel conductance from the carrier-mediated transport. The analysis of these fluctuations permits estimating the average lifetime of a channel in an open state and the conductance of a single channel. Apparently with the stochastic character of the channel opening and closing, the magnitude of its conductance (g) would be about its averaged value ($\bar{g}$) with the mean square deviation or dispersion σ^2

$$\sigma^2 = \langle (g - \bar{g})^2 \rangle.$$

It should be reminded that the channel conductance (g) for the given ion is proportional to its permeability P and concentration C

$$g \sim PC.$$

The statistical analysis of noises upon opening–closing of the channels shows that the average magnitude of the membrane conductance is related to the single channel conductance (h_0) by a simple relation

$$\sigma^2 = h_0 g.$$

Hence, we can determine the h_0 value of a single channel and its lifetime τ_0 in an open state, which may vary from milliseconds to seconds.

Active Transport. Systems of active transport provide for the transport of ions against their electrochemical potential gradient by utilizing

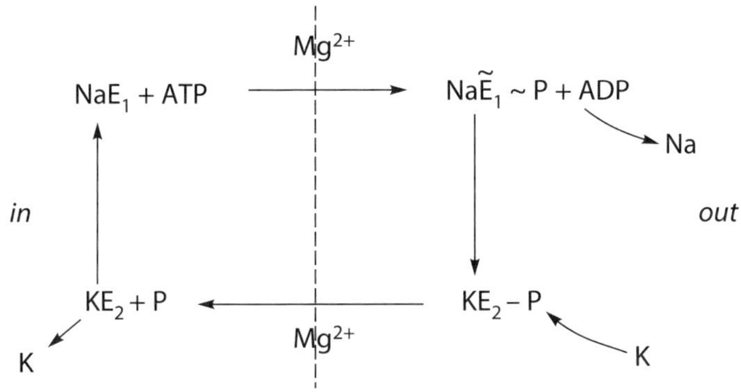

Figure 15.4 Scheme of K/Na-ATPase functioning.

the metabolic energy (ATP and coupled redox processes). Active transport in neurons is performed by the Na/K-ATPase localized in plasmatic membranes. The principle of the Na/K-ATPase action is that at the expense of the energy released upon ATP hydrolysis, it undergoes conformational changes accompanied by Na ion transfer from the cell into the extracellular medium with a concurrent movement of K^+ ions in the reverse direction. Figure 15.4 shows a scheme of functioning of Na/K-ATPase consisting of two polypeptide subunits (a small one and a large one). On the membrane interior the protein is phosphorylated in the presence of Na^+- and Mg^{2+}-ATP, the NaE_1~P complex is formed in which the protein part of enzyme E is in the strained conformation, and the complex moves to the outer side of the membrane. Then protein dephosphorylation on the outer side of the membrane alters the enzyme affinity to Na^+ and K^+ ions. Now Na^+ ions strongly bound to protein, are expelled into the external medium and the enzyme acquires a novel conformational state (KE_2) has after binding K^+ with a high affinity. After that the KE_2 complex migrates to the internal medium, where K^+ is released and the enzyme returns to the initial state E_1 to reenter the conversion cycle.

The stable phosphorylated complex $\widetilde{E}_1 - P$ binds three Na^+ ions and, when in state E_2, two K^+ ions. Thus, the ATP energy is expended to form such a complex in a strained conformation state Na $\widetilde{E}_1$~P in which the affinity to Na^+ diminishes and the affinity to K^+ increases. In this state, the substitution of 3 Na^+ for 2 K^+ relieves the strain. So, owing to the ATP energy, an ordered change in the affinity of the enzyme ion–binding centers to cations takes place. In the membrane, the shifting of subunits carrying ions is caused by thermal fluctuations. One of the models suggests that the ion-binding site of the enzyme contains a coordination sphere with

12 oxygen atoms. As known, four oxygen atoms can form a Na^+–specific sphere and six oxygen atoms a K^+–specific sphere. Rearrangement of the 12 oxygen atoms from the state where 3 Na^+ are bound (four oxygen atoms per Na^+ ion) to the state where 2 K^+ are bound (six oxygen atoms per K^+ ion) needs conformational reorganization of the protein and correspondingly the ATP energy. The Na/K-ATPase functions in an electrogenetic mode because the exchange of 3 Na^+ ions for 2 K^+ ions gives rise to an additional potential difference across the membrane.

Membranes also contain Ca^{2+}–dependent ATPase that has some features similar to those of the Na/K-ATPase. Its work cycle includes the binding of Ca^{2+} and ATP with the formation of a conformational unstable state in which the affinity of the protein to the Ca ion changes due to the high-energy phosphate bond. As in the case of the Na/K-ATPase, here the affinity change is evidently caused by the alteration in the position of polar groups binding Ca^{2+} in the coordination sphere of the enzymatic center.

Active transport of H^+ ions can occur through H^+–channels of the H^+–ATPase along which they are delivered to the active center where the ATP synthesis and hydrolysis take place.

The mechanism of the H^+ transport itself is not completely understood. It is proposed that there exists the relay race transfer of protons via the net of hydrogen bonds accompanied by mutual displacement of donor–acceptor groups. The structure of energy barriers determining the rate and direction of the H^+ transport in the H^+–channel depends on its conformational state. Therefore the ion energy profile in the channel can be modified at the expense of the energy of the ATP, light quantum (the H^+–channel in bactiorhodopsin) and redox reactions.

16

Transport of Ions in Excitable Membranes

Action Potential. The rest state is characterized by the following relation of potassium and sodium permeability $P_K : P_{Na} = 1 : 0{,}04$, i.e. $P_{Na} \ll P_K$. This explains the approaching of the resting potential values to the equilibrium potential of K^+ ions in the unexcited membrane. In the rest state, the internal content of the nerve fiber is charged negatively relative to the outer medium which is determined by the K^+ movement outwards from the axoplasm. From the Goldman equation the resting potential is

$$\varphi = \frac{RT}{F} \ln \frac{P_K[K_0] + P_{Na}[Na_0]}{P_K[K_i] + P_{Na}[Na_i]}.$$

Excitation of a nerve fiber membrane causes generation of electric pulses. A single-shot nerve impulse (the action potential) lasts for about 1ms and propagates along the fiber at the rate of 1–100 m/s. Upon generation of the action potential the values of the potential inside the fiber alter rapidly from negative −70 mV to positive +40 – +50 mV (depolarization) and then return to the initial level via a short-time stage of hyperpolarization

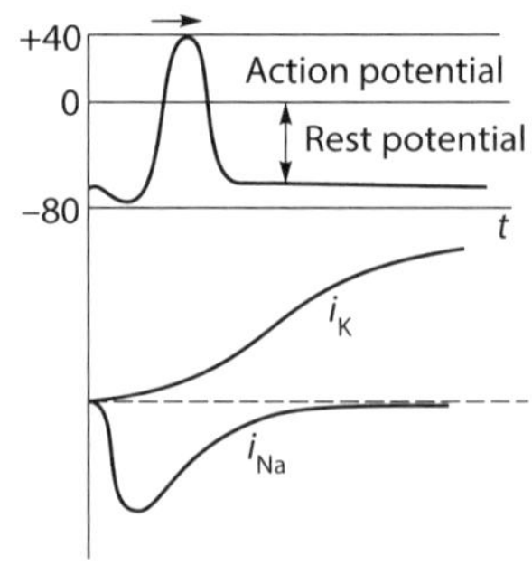

Figure 16.1 Generation of the action potential impulse as a result of local depolarization.

(Fig. 16.1). Reversion of the membrane potential during the impulse generation is caused by a sharp change in the membrane permeability for ions, so that at the spike peak

$$P_K : P_{Na} = 1{:}20, \text{ i.e. } P_{Na} \gg P_K.$$

So, the membrane permeability to ions depends on the membrane potential applied. As the spike increases, the flux of Na^+ ions directed inwards causes depolarization until the inward and outward Na^+ fluxes become equal and the membrane potential becomes close to the sodium equilibrium potential. Then Na-channels are inactivated and the inward Na^+ flux is terminated. Concurrently the permeability for K is enhanced and K^+ is now released outwards following the gradient of its electrochemical potential. During this process the membrane is repolarized until the release of K is terminated, and the membrane potential becomes close to the potassium equilibrium potential. Repolarization occurs via a fast phase of hyperpolarization. The membrane potential reaches again the level of the resting potential, but this happens under conditions of an increased concentration of sodium ions and a decreased concentration of potassium ions in the cell. The above deviations from the ion distribution on the resting membrane can be enganced the repeated propagation of nerve impulses occurs. In such conditions, the constant level of intracellular concentrations of K^+ and Na^+ is maintained by the Na/K-ATPase which removes sodium ions in the exchange for potassium ions.

Channel Conductance. Gating Currents. The change in the Na^+ and K^+ fluxes (i_{Na} and i_K) during the action potential (Fig. 16.1) is caused by two types of ion channels selective for sodium and potassium. Their conductance alters differently depending on the membrane electric potential. The sodium permeability grows rapidly and then promptly drops exponentially.

The potassium permeability increases following an S-shaped curve and in 5-6 ms becomes constant. The reduction of the sodium permeability to the initial values is 10-fold faster than that of the potassium permeability. The problem of how the permeability of ion channels is regulated by the electric field is a central one in the biophysics of membrane processes. The Hodgkin-Huxley model suggests that conductance for sodium and potassium ions is regulated by positively charged controlling particles that move in the membrane following changes in the membrane electric field. The displacement of the position of these particles in the membrane depends on the potential applied and opens or closes the ion channel accordingly. It is assumed that in the case of potassium permeability there are four particles activating the channel conductance. And in the case of a sodium channel, it is proposed that three activating particles are required for opening the channel while to close it one inactivating particle is required. Based on these assumptions a mathematical model was designed which simulates nerve impulses with a high accuracy. The major advance of the model was the separation of transmembrane currents into two components (i_{Na} and i_K) and in experimental studying their properties. Elements responsible for the ion selection mechanisms (the selective filter), activation (activation gates) and channel inactivation (inactivation gates) were determined in the functional structure of the channel (Fig. 16.2). The motion of charged

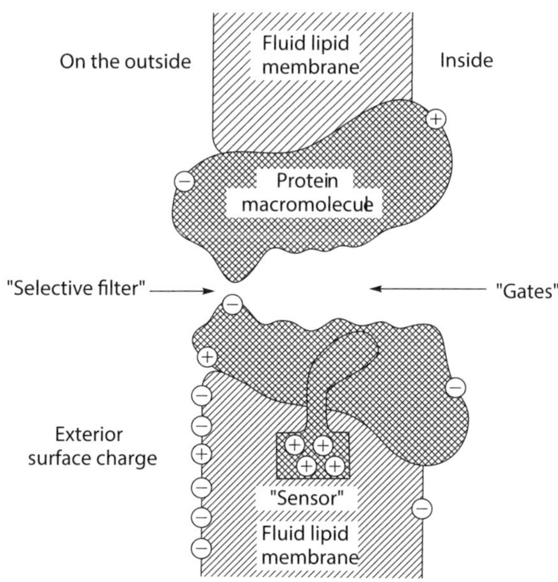

Figure 16.2 Structure of an ion channel (according to Hille, 1981).

particles (gating particles) in the channel is detected experimentally upon the generation of gating currents. They appear as a result of the displacement of particles in the membrane caused by the electric impulse applied to the membrane. Displacement-based gating currents associated with particles opening the Na-channel were discovered. At the same time the assumption that the jump of several charged groups should occur across the whole thickness of the membrane seems to be dubious. That is why general interpretations of gating currents are put forward. For example, it is suggested that displacement currents are caused not by the transmembrane jump of charged particles, but by the cooperative change in the orientation of dipoles lining the channel internal cavity.

The polar groups shaping the dipole environment of the ion in the channel affect the ion energy and its transport through the channel. If the depolarizing electric impulse changes the orientation of the polar groups, this causes a displacement of the charged groups and alters the channel conductance. The process of dipole reorientation can have a cooperative character and be fairly abrupt. In this case the energy required for reorientation of every elementary dipole should depend not only on its own energy, but also on the portion of dipoles that have already changed their orientation. In other words, as the orientation of a portion of dipoles changes, the energy necessary for the reorientation of the remaining part of dipoles decreases with the increase in the number of the already reoriented dipoles. It can be assumed that at the beginning the dipoles interfere with each other and prevent changing the orientation under the action of the field, while the already reoriented dipoles no longer block the remaining ones. As a result of such a cooperative effect, the channel conductance grows rapidly in an avalanche-like manner under the influence of the electric impulse applied. Modern literature data on channel structural rearrangements coupled to sharp close-open state transitions in a single channel illustrate phenomenological dipole orientation models which may include negatively charged amino acid residues in a channel.

Nerve Impulse Propagation proceeds without damping along a fiber. At the site of the spike, the internal part of the fiber has a positive charge while in the adjacent resting sites it is negative (Fig. 16.3). Consequently a local current is generated between the excited and resting sites. Its direction is such that it depolarizes the membrane region in the vicinity of the active site which also results in its excitation and spiking. Thus, the excitation travels further across the fiber. A nerve fiber can be compared to an electric cable the adjacent sites of which are interconnected. But in contrast to a standard cable, propagation of the nerve impulse is not accompanied by its attenuation because the membrane contains molecular "generators"

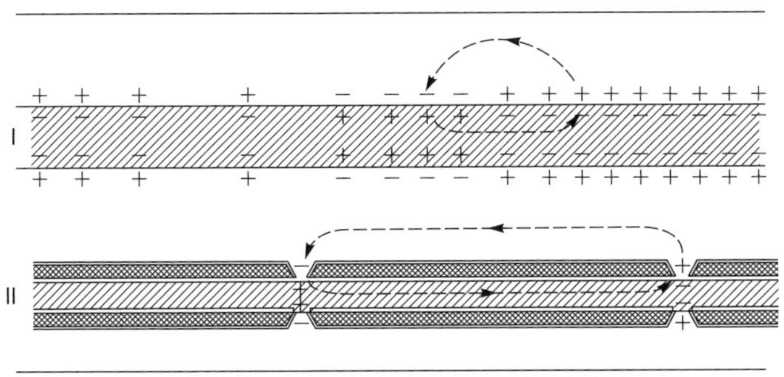

Figure 16.3 Charging of the nerve fiber surface upon impulse propagation. (Illustration of the Hodgkin theory of local currents, 1965)

injecting the traveling impulse. Determination of the rate of propagation of a nerve impulse calls for studying the kinetics of ionic transmembrane currents (i_{K+}, i_{Na}) and the leakage current as well as the knowledge of electrical properties of nerve fibers (capacitance and resistance). The analysis of corresponding mathematical models shows that the propagation rate of a stable impulse is inversely proportional to a square root of the fiber diameter, which is supported by the experimental data. In organisms nerve fibers are associated in nerve bundles where each fiber can independently conduct the excitation. The electric field generated in the excited fiber affects the membrane potential of adjacent fibers, which under certain conditions can thereby be excited too.

17

Primary Processes of Energy Transformation in Photosynthesis

The sequence of separate reactions in photobiological processes usually includes the following stages: absorption of a light quantum by a chromophore group and generation of electronic excited states $\rightarrow$ transfer of electronic excitation energy $\rightarrow$ primary photosynthetic act and generation of primary photoproducts $\rightarrow$ formation of primary stable chemical compounds $\rightarrow$ physiologobiochemical processes $\rightarrow$ final photobiological effect.

One of the tasks of biophysics is to elucidate the mechanisms of initial stages following directly the absorption of a light quantum. As will be shown, in spite of a great diversity of photobiological effects, the initial stages of light energy transformation have common molecular mechanisms. An intricate combination of redox reaction necessary for the overall electron transport along the electron transport chain (ETC) underlies the primary processes of photosynthesis.

General Scheme. Figure 17.1 shows a standard Z-scheme of primary processes of photosynthesis in higher plants with times of separate electron transport steps. As seen, photosystems PS I and PS II function in series: photoproducts reduced by PS II serve as electron donors for PS I. Far-red light ($\lambda > 680$ nm) is absorbed mainly by pigments of PS I and

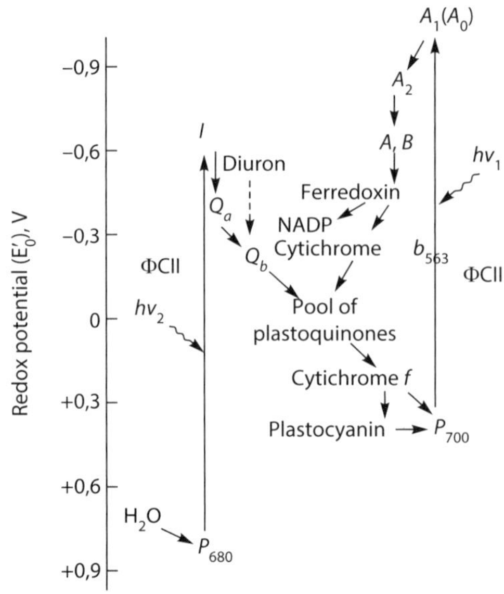

Figure 17.1 Scheme of primary processes of photosynthesis in higher plants. In compliance with the concept on successive interaction of two photosystems in the so-called Z-scheme of photosynthesis, the reduced products of PS II served as electron donors for PS I. Actinic light which is mostly absorbed by PS II results in the reduction of intermediate carriers in the ETC, while, on the contrary, PS I excitation leads to their oxidation.

causes oxidation of cytochrome which is reduced by PS II at absorption of short-wave light ($\lambda < 680$ nm). The optimal intensity of photosynthesis is observed at a certain relationship between the numbers of excited PS I and PS II which is dependent on the spectrum composition of light illumination. Light is absorbed by pigments of the light-harvesting (LH) pigment–protein complex (LHPPC) which, like a funnel reservoir, transmits the excitation energy to the pigment–protein complexes of PS I and PS II (PPC1 and PPC2) and then directly to reaction centers RC1 and RC2 (Fig. 17.2). Depending on the conformational state of photosynthetic membranes, the topography of the PPC1 and PPC2 arrangement as well as their connection with LHPPC change, thus affecting the distribution of excitation energy between PS I and PS II. This is governed by the presence of ions in the medium, pH of the medium, level of phosphorylation and the surface charge of proteins in LHPPC. Under different physiological conditions, the role of these factors varies, which therefore is of a regulatory significance for the excitation energy distribution between the photosystems.

The measure of the efficient deactivation of primary electron excitation in photosynthesis is a decrease in the lifetime of the excited state of chlorophyll.

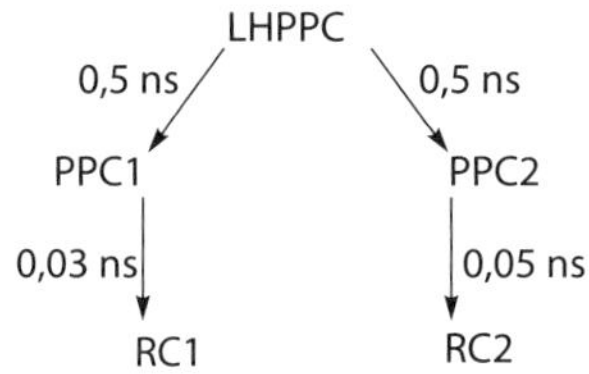

Figure 17.2 Scheme of energy migration in the photosynthetic apparatus of higher plants.

Measurements of the (τ) of the chlorophyll singlet excited state (S_1) lifetime showed that the τ value in chlorophyll solution makes 5 ns whereas in chloroplasts it decreases by 20 to 30 times. Evidently, when excitation is delivered to the RC, an efficient process of utilization of the electronic excitation energy S_1 state occurs in the primary act of photosynthesis. Thus the time (τ) of the chlorophyll excited state (S_1) in a green leaf diminishes. The quenching of fluorescence of various pigment forms has a complicated character. The decrease in the lifetime of the excited state of pigments to 0.5 ns takes place in LHPPC, which delivers the excitation energy to PPC1 and PPC2. And only then more efficient quenching of excitation proceeds in RC1 and RC2 during 0.03–0.05 ns (Fig. 17.2). Energy transfer among separate complexes (LHPPC, PPC1 and PPC2) proceeds by the inductive resonance mechanisms. Within one PPC it has an exciton character (Chapter 11).

The primary act of photosynthesis in RC occurs when excitation from a PPC is delivered to a photoactive pigment (P) of the RC itself, an electron is released and the primary acceptor (A_1) is reduced. After that the acceptor delivers an electron to the next acceptor (A_2) and so forth in the electron-transport chain:

$$(PPC)^* + PA_1A_2 \rightarrow P^*A_1A_2 \xrightarrow{\leq 10^{-11}\ s} P^+A^-_1A_2 \xrightarrow{0.1\ ns} P^+A_1A_2^- . \qquad (17.1)$$

P^+ is then again reduced from electron donors in the electron-transport chain. Derivatives of porphyrin molecules (for example, pheophetin in RC2) play the role of A_1 in RC, while the role of the secondary acceptor is played by quinone molecules (RC of purple bacteria) and plastoquinone (RC2), or by iron-sulphur proteins (RC1). The efficiency of the electron release from P^+ and its delivery to A_1 varies from 95 to 98% and takes less than several picoseconds (1 ps = 10^{-12} s). It should be noted that, if the secondary acceptor A_2 in the reactor center has been chemically reduced in the dark, then the light induced transfer of the electron to A_1 brings about its recombination with P^+ in 5–10 ns generating P^*. This recombination is accompanied by the delayed luminescence

$$P^*A_1A^-_2 \xrightarrow{<10\text{ ns}} P^+A_1^-A_2 \xrightarrow{5\text{-}10\text{ ns}} P^*A_1A_2^- \rightarrow PA_1A_2^- + h\nu.$$

The kinetics of electron transport between P (the bateriochlorophyll dimer), A_1 (bacteriopheophetin, BPP), A_2 (primary quinone, Q_A) and A_3 (secondary quinone, Q_B) in the reaction center of purple bacteria was studied in detail.

The scheme of the electron transport looks like this

$$PB_{PP}Q_AQ_B \xrightarrow{h\nu} P^*B_{PP}Q_AQ_B \rightarrow P^+\acute{A}^{-\bullet}_{\hat{o}\hat{o}}Q_AQ_B \xrightarrow{150\text{–}200\text{ ns}}$$

$$P^+B_{PP}Q_AQ_A^{-\bullet} \xrightarrow{100\,\mu s} PB_{PP}Q_AQ_B^{-\bullet} \xrightarrow{2\,\mu s} PB_{PP}Q_AQ_B^{-\bullet} \quad (17.2)$$

In these experiments, excitation was accomplished by sharp laser pulses (1–2 ns) which allowed studying the kinetics of slower electron transport stages. The general scheme of primary bacterial photosynthetic processes is as follows:

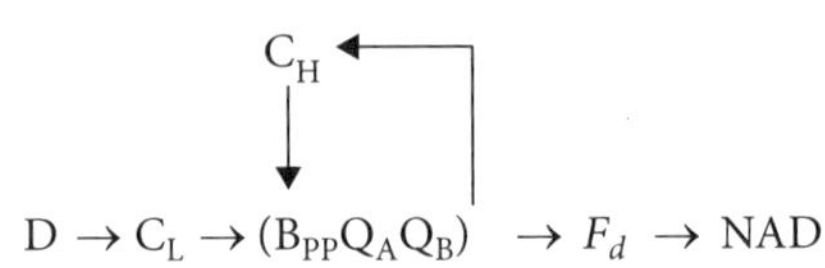

(17.3)

where hydrogen sulfite and organic molecules are an external source of electrons for the reduction of NAD vriaFerredoxin. Cytochrome C_H provides for a cyclic electron flow and cytochrome C_L ensures the relationship of the cyclic transport with external electron donors.

Physical Mechanisms of Electron Transport. The most important property of RC functioning is that individual stages of electron transport can proceed efficiently also at low temperatures down to liquid-nitrogen and liquid-helium temperatures. Figure 11.5 demonstrates a typical two-phase curve for temperature dependence of photooxidation of "noncyclic" cytochrome C_L by the photoactive pigment P in the reaction center of purple bacteria. The scheme of this reaction looks like this

$$C_LP^*A_1A_2 \xrightarrow{<10\text{ ps}} C_LP^+A_1^-A_2 \xrightarrow{\sim 1\,\mu s} C_LP^+A_1A_2^- \rightarrow$$

$$\xrightarrow{\sim 1\,\mu s} C_L^+PA_1A_2^-.$$

Apparently the tunneling mechanism provides for the oxidation reaction of C_L at low temperature.

As has been described (see Chapter 11), the initial part of the curve in Fig. 11.6 refers to the temperature-dependent activation processes of nuclei excitation and their transition on higher vibrational levels where the donor–acceptor complex tunnels to the final state. At the same time the tunneling process requires preliminary formation of a contact state between the donor and acceptor groups. This is achieved because of the intramolecular mobility of the protein in the reaction center where prosthetic groups of carriers are located. Hence the formation of contact states also contributes to the temperature dependence of the electron transport. The temperature dependence of the electron transport should correlate with the intramolecular mobility of the reaction center protein as demonstrated in experiments. Figure 17.3 shows temperature dependencies of the direct electron transport between primary and secondary quinones and also of the characteristic time τ' of the spin label mobility attached to the reaction center protein. It is seen that, when the temperature of samples rises from 140–180 K, τ' drops sharply which is the evidence of "defrosting" of intramolecular motions in the reaction center protein.

It is just in this temperature range that functional activity of the reaction center resulting in the electron transport to secondary quinone also increases. Analogous correlation is revealed in experiments in which

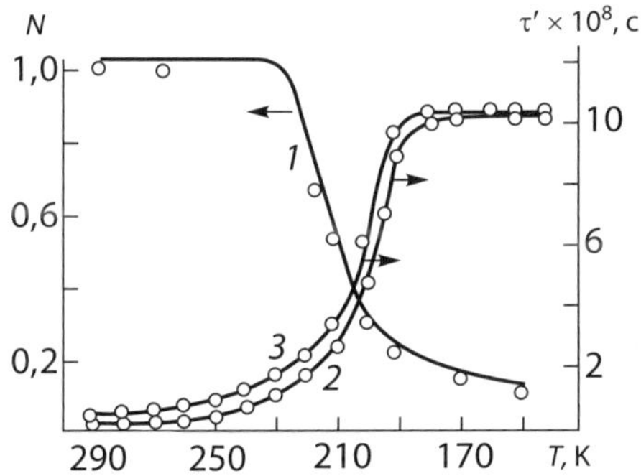

Figure 17.3 Dependence of functional activity (1) and conformational dynamics (2, 3) of reaction centers in spin-labeled chromophores of *R.rubrum* versus temperature. N is the efficiency of photoinduced electron transport from Q_A to Q_B (curve 1); τ' is the effective parameter of time correlation of rotatory diffusion of a hydrophobic spin probe (curve 2) and a spin label on SH-group (curve 3) (А.И.Берг, П.П.Нокс, А.А.Фролов, И.Н. Храмов, А.Б.Рубин, Г.И.Лихтенштейн, В.И.Гольданский, Ф.Парак, М.Букл, Р.Мессбауэр. Мол. биолог., 1979).

the probability of recoilless absorption of γ-quantum is measured (see Fig. 9.6). Defrosting of the protein macromolecule proceeds concurrently with its activated intramolecular motions over conformational substates (see Fig. 10.1) with amplitudes of about 0.3 Å.

At the same time as the reaction center is dehydrated, the efficiency of electron transport goes down and the protein intramolecular mobility becomes slower. The role of the formation of contact states in the system of quinone acceptors can be simplistically illustrated as follows. Having accepted the electron, primary quinone changes its electronic state and switches to another conformation coordinate which should correspond now to its reduced state (see Fig. 11.8). Here, after having reached a certain contact state with the secondary quinone, the electron is transferred to the latter. At low temperatures, the primary quinone mobility decreases and consequently the efficiency of the electron transport to the secondary quinone also goes down. However if the sample is cooled under intense light, then due to the primary electron release from P the molecules of primary quinone are also reduced, but the whole situation changes drastically. In fact, as demonstrated by experiments, under light cooling primary quinone is reduced already at the initial moments of cooling and thus it manages to switch to a contact state, where an electron transfers by the tunneling mechanism to the secondary quinone at low temperatures. Therefore in such samples cooled to low temperatures under continuous illumination, the efficiency of electron transport to the secondary quinine does not actually depend on temperature. Evidently in this case electron tunneling proceeds efficiently from low vibrational levels coupled to fast energy dissipation on the accepted mode at low temperatures. Hence it suggests that in the already "prepared" contact states, temperature dependence of the electron transport per se between carriers is weak. Indeed, the primary reduction of P_{PP} in the reaction center of purple bacteria occurring during time $\tau \leq 1.0$ ns ($P^{*}P_{PP} \rightarrow P^{+}P_{PP}^{-}$) is actually independent of temperature down to −196°. When the temperature dependence of cytochrome C oxidation involved in the cyclic flux in the reaction center of purple bacteria was studied (Fig. 17.3), it was found that in contrast to "noncyclic" cytochrome C_L, the rate of oxidation of C_H ($C_HP^{+} \rightarrow C_H^{+}P$) is in fact independent of temperature. However at temperatures below a critical one, the number of cytochrome C molecules undergoing photooxidation diminishes because of the decreased number of contact states (C_HP). The oxidation rate constant for C_H molecules, that remained active, is not changed. Disintegration of contact states (C_HP) takes place not only at a temperature decrease but also upon dehydration of the protein from the reaction center. The role of water and temperature is revealed both in supporting mobility and in stabilizing the contacts formed between donor–acceptor groups.

Interesting results have been obtained on molecular mechanisms of phototransformations of rhodopsin and its related pigment bacteriorhodopsin (Br) discovered in purple membranes of halophilic bacteria. Each bacteriorhodopsin molecule contains one chromophore – retinal (polyene aldehyde) in a complex with the opsin protein

$$\text{Retinal}-\underset{\displaystyle H}{\underset{|}{C}}=\underset{\displaystyle H}{\underset{|}{N^{+}}}-\text{Opsin}$$

The light energy absorbed by retinal is used for an active proton transport across the membrane and for the formation of the electrochemical proton gradient and ATP synthesis. The initial stage of light energy transformation takes <10 ps. It includes trans- and cysisomerization of retinal, displacement of the Schiff base proton and structure-and-polarization changes in the closest protein environment of retinal (Fig. 17.4). This triggers a cascade of sequential conformational rearrangements in the macromolecular complex retinal – opsin, which finally lead to proton translocation across the membrane, probably along the hydrogen bonding net. In the case of rhodopsin, the chromophore is also a retinal molecule linked to opsin via a Schiff base. At the beginning, retinal is in the 11-cys state. The initial act of rhodopsin phototransformations includes cys- and transisomerization of retinal coupled to the displacement of the Schiff base proton. In both cases the initial stage of photoisomerization of retinal in complex with opsin occurs much faster (less than picoseconds) than that of free retinal in solution (several nanoseconds). This fact illustrates the

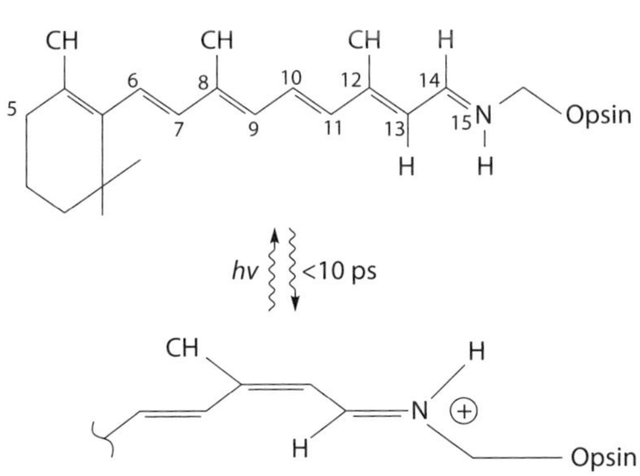

Figure 17.4 Scheme of the primary action of the bacteriorodopsin photocycle.

role of the specific protein environment of retinal which provides for not only its fast isomerization, but also stabilizes intermediate conformational states involved in the subsequent directed changes. The final result of rhodopsin phototransformations is the generation of a photoelectric signal in a receptor cell membrane of the retina. Enzyme systems take an active part in receptor response in the rod cell. It can be seen that the primary stages of light energy transformation in photoenergetic and photoinformational processes occur with the involvement of the protein environment of the chromophore. They are realized according to the principle of directed photoinduced electron-conformational interactions corresponding to specific structural changes of a photosensitive chromoproteid.

18

Energy Transformation in Biological Membranes

An important function of biological membranes is to ensure energy transduction related to its transformation from one type of energy to another. In effect this is the basis of bioenergetic processes in a cell. As known, the energy required for different kinds of life activities within the cell is utilized as the energy of the chemical bonds in an ATP molecule, which is formed mostly in biological membranes of mitochondria and chloroplasts (chromatophores). In such systems, the electron flux generated both in mitochondria due to substrate oxidation and in chloroplasts due to light energy is coupled to proton translocation as the driving force of ATP synthesis.

Different mechanisms of proton translocation across a membrane coupled to electron transport proceeding along the chain of carriers positioned asymmetrically exist in a membrane. It is probable that in the protein part of the electron carrier a proton channel is formed, its conductance being dependent on the redox state of the carrier itself. So, an electron carrier functions simultaneously as a proton pump. Proton translocation can occur also as a result of the work performed by mobile carriers which diffuse through the membrane from one side to the other. These carriers are reduced on one side of the membrane and accept both the proton and the electron. Then after diffusion to the other side of a membrane, the carriers are oxidized and expel a proton into the membrane cavity. It is in this way that such functions are performed in thylakoid membranes in chloroplasts, where it directs electrons and protons from the outer (negative) membrane surface to the internal (positive) one. It should be noted that membrane polarity in mitochondria is inverse to that of thylakoid (plus being on the outer side and minus on the internal side). Within thylakoid, protons are consumed from the outer phase and are transported to the internal phase which is thus acidified upon functioning of the electron transport chain. This results in the formation of the transmembrane concentration gradient of protons (ΔpH) between the outer and internal phases of thylakoid. Simultaneously transmembrane electric potential difference (Δφ) is generated because of the growth of the positive charge within thylakoid when positively charged protons are accumulated therein. Upon functioning of photosystems I and II, photochemical electron transport per se to the outer side also contributes to the electric potential difference across a membrane. Thus the electric field (Δφ) generated in such a way affects transport of other ions penetrating through the membrane, which in their turn change Δφ and affect proton transport. As a consequence, we have a complicated situation of mutual influence of the two components ΔpH and Δφ of the transmembrane electrochemical potential. Under steady-state conditions, the value of the electrochemical gradient on the membrane will depend on the relation of the rates of the transmembrane electrogenic ion transport, proton translocation and electron flux along the ETC.

According to the Mitchell principle of chemiosmotic coupling, electron transport is coupled to the ATP synthesis through the formation of this transmembrane difference of electrochemical potentials for protons:

$$\Delta\bar{\mu}_{H+} = F\Delta\varphi + 2.3RT\Delta pH,$$

where F is the Faraday constant.

Energy $\Delta\bar{\mu}_{H+}$ is used to synthesize ATP by a specialized enzyme, i.e. membrane ATP-synthase. It should be understood that the value of the electrochemical potential in itself characterizes the thermodynamic driving force of ATP synthesis but does not reveal the molecular mechanisms of this process.

The ATPase Complex. Proton ATPases have been isolated from all types of coupling membranes: mitochondria, chloroplasts, and chromatophores.

The ATP-synthase complex, or H^+-ATPase, is a reversible enzyme having both the ATP- synthase and the ATPase functions. The ATP synthesis is performed because of the energy $\Delta\bar{\mu}_{H+}$, and ATP hydrolysis results in coupled generation of the transmembrane difference of H^+ electrochemical potentials by proton ATPase. Therefore, in the H^+-ATPase, processes occur according to the common scheme

$$\Delta\bar{\mu}_{H+} \underset{\text{ATPase}}{\overset{\text{ATP-synthase}}{\rightleftarrows}} \text{ATP}.$$

The ATP-synthase complex (F_0F_1–ATPase) consists of soluble ATPase (factor F_1) and membrane components (complex F_0). The general topography of the ATPase complex is shown in Fig. 18.1.

The coupling factor of ATPase (factor F_1) is a polyfunctional protein with a complicated quaternary structure. It consists of three types of large subunits (α, β, and γ with a molecular mass of 30000-60000 Da) and two types of minor subunits.

F_0 forms a specifically organized channel that allows a proton to pass through the membrane from the aqueous phase to the hydrophobic region of the membrane and then from the latter to water on the other side of the lipoprotein barrier. The most probable mechanism of proton transport

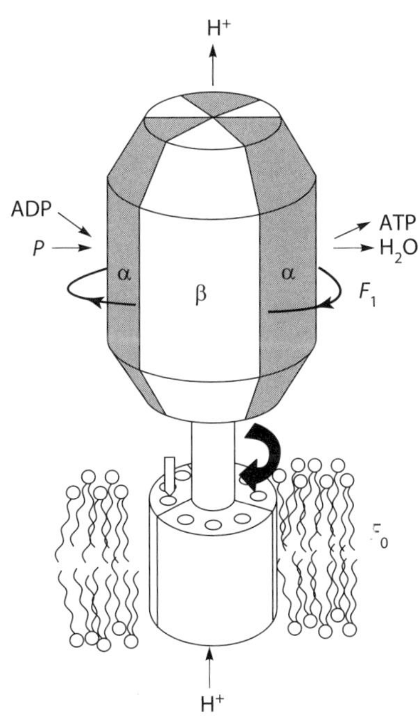

Figure 18.1 Simplified scheme of the H^+–ATPase complex (according to Lanyi, Nature, 1995).

seems to be a proton hopping along proton-donor and proton-acceptor groups of amino acids including Arg, Tyr and Glu.

The basic catalytic properties of F_0 factor are related to α- and β-subunits where the binding sites of nucleotides with varying affinity are located. The α-subunit of the enzyme has non-catalytic centers possessing high affinity to ADP and high specificity of its binding.

The problem of ATP synthesis in the H^+–ATP-synthase should be considered based on general concepts on the role of electron–conformational interactions in the mechanisms of enzyme catalysis as discussed above.

As known, the elementary act of catalysis occurs in the active center of an enzyme spontaneously, when it acquires an active configuration of the reacting substrate groups and those of the enzyme active center, positioned at distances of about the lengths of chemical bonds. At the stages when the substrate and enzyme interact during the formation of their active configuration and then upon the release of the formed product the enzyme undergoes conformational changes. Such external factors as temperature, ionic strength and viscosity can influence these relaxation stages. However, the direct act of catalysis in the formed active configuration does not require thermal activation energy.

So, the energy $\Delta\bar{\mu}_{H+}$ is used mainly to expel strongly bound synthesized ATP from the catalytic center of the enzyme. The energy is also expended during the binding of phosphate and ADP by the enzyme provided precise functioning of the catalytic centers. This means that the energy $\Delta\bar{\mu}_{H+}$ is expended not at the stages of elementary formation of a covalent ADP–P_H bond in the enzyme active center but rather during both the substrate binding by the enzyme and release of the reaction products from the active center (ATP or ADP and P_H).

The complex has active centers that can be either in the open or closed states. When ATP is synthesized, the binding of ADP and P_H in one center evokes such a change of the conformation, which contributes to the release of the earlier formed ATP from the other center. Similarly, in the process of ATP hydrolysis the binding and hydrolysis of ATP result in rapid dissociation of ADP and P_H form the alternative center. The role of the $\Delta\bar{\mu}_{H+}$ components is to protonate certain molecular groups or to change the position of polar groups by the action of the electric field $\Delta\varphi$, which alters the character of motion of individual protein groups in the conformational potential. This may directly affect both the formation of the enzyme–substrate complex during the substrate binding and the detachment of the reaction product, i.e. the processes occurring in the course of conformational relaxation.

It was demonstrated that during ATP hydrolysis the γ-subunit can rotate around its axis revolving relative to the coupling factor F_1 "sitting" on the axis, due to the energy of ATP hydrolysis. Using a fluorescent microscope, researchers from Japan observed motions of the γ-subunit during ATP hydrolysis by F_1 molecules. Three β-subunits of the F_1 complex were immobilized on a fixed substrate and a fluorescent fragment of an actin filament of 1 μm long was chemically cross-linked to the terminus of the γ-subunit. During the enzyme functioning and ATP hydrolysis, rotation of the actin filament took place which terminated in the absence of ATP.

At each turn, the rotation of the γ-subunits in the F_1-complex moves its β-subunits to a new position where they appear in different microenvironment because of some asymmetry. This is the reason for consecutive changes in the state and conformational rearrangement of active centers in the F_1 complex.

During ATP synthesis (Fig. 18.2), the states of β-subunit centers in the F_1-complex change. In state 1 the center of the β-subunit is open and its ATP and P_H molecules, comparatively loosely retained by the center, are weakly linked. Rotation of γ-subunit and conformational rearrangements of the F_1 complex transfer this center to state 2 where the ATP synthesis per se takes place. Here ADP and P_H are strongly fixed in the catalytic center in active configuration necessary for the formation of a covalent bond between the phosphate groups of ADP and P_H. That is why, as mentioned above, at this stage no inflow of energy from the outside is needed. In state 2 the formation and splitting of the covalent ADP–P_H bond occur

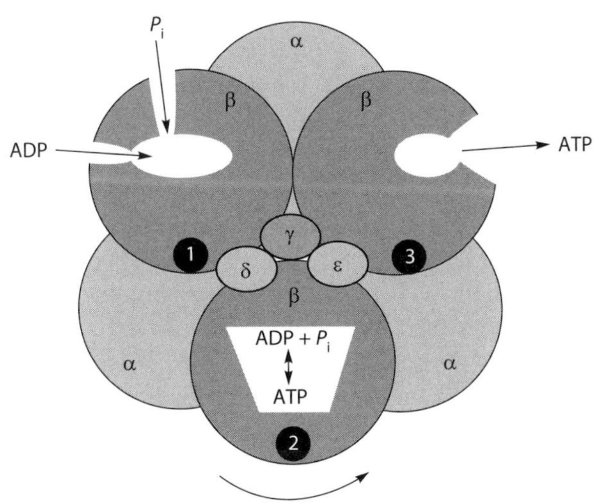

Figure 18.2 Scheme demonstrating energy-dependent changes in the state of catalytic centers.

spontaneously because the equilibrium constant of the reaction ADP → ADP + P_H is close to unity.

At the next stage of rotation, the center switches to state 3 in which, due to the energy-dependent structural rearrangement, the strong link of the ATP molecule with the center is attenuated and the molecule is released to the outside. The empty place is occupied by new ADP and P_H molecules. The described rearrangements have a cooperative character and affect the state of all the three catalytic centers of β-subunits in complex F_1. Since the total cycle involves three stages and the H^+-ATPase has three subunits, each structural transition leads to the appearance of a new ATP molecule.

The rotation is stepwise and is accompanied by discrete turns at a definite angle each. This conclusion was substantiated by experimental results of high-speed shooting of rotation of γ-subunits with an attached actin fiber during hydrolysis of ATP.

So, upon functioning of the ATPase complex, chemical energy of ATP hydrolysis/synthesis and transmembrane gradient of protons are mutually transformed into mechanical work of the rotation of protein subunits.

Such transformation of chemical and mechanical energy occurs also in other proteins performing motor functions. Examples are kinesin directed motions along a microtubule, DNA transport by bacteriophage portal protein, flagella motions and directed motions of DNA polymerase along a DNA strand.

In all these cases, on account of their small energetic and spatial scales distinct displacements of protein parts of the motor are essentially affected by the Brownian movement of the environment, i.e. water molecules and other atomic groups of proteins. Moreover, the Brownian movement itself and thermal fluctuations are necessary for the motor functioning. This is the difference of molecular motors from macroscopic motors, which are in general affected by temperature but not by thermal fluctuations.

Directional structural rearrangements in the motor are a consequence of its specific molecular configuration and its alterations with the changes in the protein electric field.

An example of such mechanochemical coupling in protein is chemical binding of a substrate molecule in the enzyme active center, which causes directed conformational rearrangements in a protein. Nondirectional mechanical displacements of a protein occur spontaneously due to the stochastic Brownian forces, but only coupling to a chemical reaction that changes the electron-conformational equilibrium, imparts a directed character to these displacements. This principle is the basis for a molecular motor model, a specific mechanism identified as the Brownian ratchet. As known, in a ratchet the gear wheel rotates only in one direction under the action of an external force. Its rotation in the reverse direction is prevented by a "locking pawl" which slides under one of the teeth. In the Brownian ratchet, the energy source for the motor movement is thermal fluctuations which can, in principle, cause a displacement of the protein subunit in two opposite directions. Due to the action of the thermal fluctuations, the motor can switch to one of the two possible states, one of which is "necessary" and the other "parasitic". The chemical reaction, the products of which are disabling the "parasitic" state at every step, plays the role of a ratchet ("locking pawl") (Fig. 18.3). This happens upon dissociation of proton-containing groups on the rod in the left-hand cell at high pH. As a result the rod acquired a negative charge and thus cannot move now to the right-hand cell across the dividing hydrophobic membrane. In the right-hand cell the rod is neutral at low pH and can move to the left-hand cell due to thermal fluctuations.

Given the frequency of thermal fluctuations lower than that of chemical acts, at each step the blocking has time to happen which results in the transition of the rod from the right-hand to the left-hand cell. In the model this corresponds to the protein transition only to the necessary state (Oster, 2000).

So, because of the chemical work only fluctuations which act in the "required" direction are "selected" and thereby the process stochastic by its nature acquires a vector character. Naturally, this does not violate the second law of thermodynamics since this selection is provided by an independent source of chemical energy due to the initially generated difference in pH values in the two parts of the cell.

In the ATP-synthesis the internal part (the cytoplasm) has a low pH while in the external part (the periplasm) the pH is high. The complex

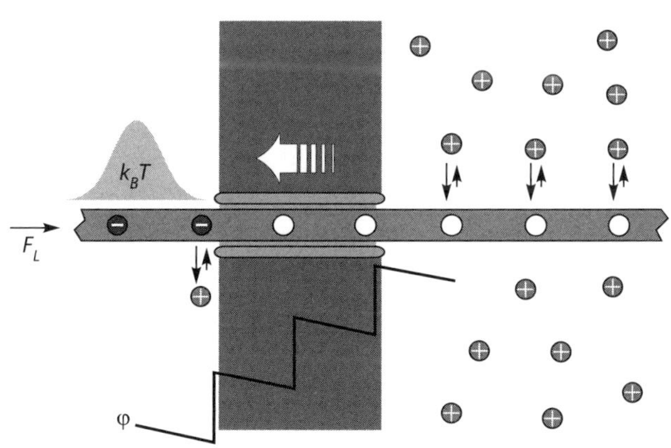

Figure 18.3 The Brownian ratchet.

operates in the mode of ATP-synthesis because of the energy of the transmembrane gradient of protons. There exist two proton water semichannels (the right- and left-hand ones) enabling the proton transport between the internal and outer regions separated by a hydrophobic cell membrane. Residues Asp61 carrying deprotonated negative charges are located on the subunits of the rotating part of the ATP-synthase.

In a simplified form, the motor functions as follows. Thanks to thermal fluctuations, the protonated Asp61 residues can get into the semichannels. In the right-hand semichannel, Asp61 would retain its protonated state because of the contact with the low-pH region in the cytoplasm. In the left-hand semichannel, Asp61 would be negatively charged because of dissociation of the proton to the high-pH region in the periplasm.

When the neutral protonated Asp61 residue moves to the right, its place in the semichannel is occupied by the negatively charged $Asp61^{-}$ residue from the left-hand semichannel which cannot move into the hydrophobic region. Another Asp61 residue enters the left-hand semichannel and is deprotonated there once again. In this mode the common displacement to the right takes place, i.e. one step of the rotation cycle is accomplished.

Further Reading

1. Andrew Rubin, Gallina Riznichenko. Mathematical Biophysics. Springer, N. Y., 2014.
2. Professor Meyer B. Jackson. Molecular and Cellular Biophysics. Cambridge University Press, 2006, New York. Information on this title: www.cambridge.org/9780521624411.
3. Hans V. Westerhoff, Karel Van Dam. Thermodynamics and control of biological free-energy transduction. Elsevier, 1987.
4. Ilya Prigogine, Dilip Kondepudi. Modern Thermodinamics. From Heat Engines to Dissipative Structure. John Wiley & Sons, 1998.
5. Charles R. Cantor, Paul R. Schimmel. Biophysical chemistry. 1980, V. 1–3. W. H. Freeman, 1980.
6. A. Y. Grosberg, A. R. Khokhlov. Giant Molecules: Here, There, and Everywhere, 2-nd Ed. World Scientific Publishing Company, 2010
7. J.D. Murray. Mathematical Biology. Springer. Mathematical Biology. 3rd edition in 2 volumes: Mathematical Biology: I. An Introduction (551 pages) 2002; Mathematical Biology: II. Spatial Models and Biomedical Applications (811 pages) 2003 (second printings 2004).
8. Riznichenko G.Y. Lectures on mathematical models in biology (in Russian). RCD, Moscow-Izhevsk, 2011.
9. Igor N. Serdyuk, Nathan R. Zaccai, Joseph Zaccai. Methods in Molecular Biophysics: Structure, Dynamics, Function. Cambridge University Press, Cambridge, 2007.
10. L. A. Blumenfeld. Problems of Biological Physics. Springer Verlag, Berlin, 1981.
11. Robert B. Gennis. Biomembranes: Molecular Structure and Function. Springer Verlag, 1989.
12. A. V. Finkelstein, O. B. Ptitsyn. Protein physics. A course of lectures. Academic, Amsterdam, 2002.
13. M. D. Frank-Kamenetskii. Unravelling DNA: The Most Important Molecule of Life. VCH Publishers (UK) Ltd., Cambridge, 1993; (paperback version of the second edition was published in 1997).
14. Govindzhi. Photosynthesis (in Russian). Academic Press, 1982 (Govindzhi, red., Fotosintez, tom 1, 2, Mir, Moskva, 1987).
15. B. Hille. Ionic Channels of Excitable membranes. Sinauer Associates Inc., 1992.

Index

Also of Interest

Check out these published and forthcoming related titles from Scrivener Publishing

Smart Membranes and Sensors
Synthesis, Characterization, and Applications
Edited by Annarosa Gugliuzza
Published 2014. ISBN 978-1-118-42379-0

Advanced Biomaterials an-d Biodevices
Edited by Ashutosh Tiwari and Anis N. Nordin
Published 2014. ISBN 978-1-118-77363-5

Biosensors Nanotechnology
Edited by Ashutosh Tiwari and Anthony P. F. Turner
Published 2014 ISBN 978-1-118-77351-2

Advanced Sensor and Detection Materials
Edited by Ashutosh Tiwari and Mustafa M. Demir
Published 2014. ISBN 978-1-118-77348-2

Advanced Healthcare Materials
Edited by Ashutosh Tiwari
Published 2014 ISBN 978-1-118-77359-8

Advanced Carbon Materials and Technology
Edited by Ashutosh Tiwari and S.K. Shukla
Published 2014. ISBN 978-1-118-68623-2

Responsive Materials and Methods
State-of-the-Art Stimuli-Responsive Materials and Their Applications
Edited by Ashutosh Tiwari and Hisatoshi Kobayashi
Published 2013. ISBN 978-1-118-68622-5

Nanomaterials in Drug Delivery, Imaging, and Tissue Engineering
Edited by Ashutosh Tiwari and Atul Tiwari
Published 2013. ISBN 978-1-118-29032-3

Biomimetics
Advancing Nanobiomaterials and Tissue Engineering
Edited by Murugan Ramalingam, Xiumei Wang, Guoping Chen, Peter Ma, and Fu-Zhai Cui
Published 2013. ISBN 9781118469620

Encapsulation Nanotechnologies
Edited by Vikas Mittal
Published 2013. ISBN 978-1-118-34455-2

Advances in Contact Angle, Wettability and Adhesion
Volume 1
Edited by K.L. Mittal
Published 2013. ISBN 978-1-118-47292-7

Biomedical Materials and Diagnostic Devices Devices
Edited by Ashutosh Tiwari, Murugan Ramalingam, Hisatoshi Kobayashi and Anthony P.F. Turner
Published 2012. ISBN 978-1-118-03014-1

Intelligent Nanomaterials
Processes, Properties, and Applications
Edited by Ashutosh Tiwari Ajay K. Mishra, Hisatoshi Kobayashi and Anthony P.F. Turner
Published 2012. ISBN 978-0-470-93879-9

Integrated Biomaterials for Biomedical Technology
Edited by Murugan Ramalingam, Ashutosh Tiwari, Seeram Ramakrishna and Hisatoshi Kobayashi
Published 2012. ISBN 978-1-118-42385-1

Integrated Biomaterials in Tissue Engineering
Edited by Murugan Ramalingam, Ziyad Haidar, Seeram Ramakrishna, Hisatoshi Kobayashi, and Youssef Haikel
Published 2012. ISBN 978-1-118-31198-1

The Physics of Micropdroplets
Jean Berthier and Kenneth Brakke
Published 2012. ISBN 978-0-470-93880-0

Introduction to Surface Engineering and Functionally Engineered Materials
By Peter Martin
Published 2011 ISBN 978-0-470-63927-6